Using Animal Models in Biomedical Research

A Primer for the Investigator

Using Animal Models in Biomedical Research

A Primer for the Investigator

Pierce K. H. Chow
Robert T. H. Ng
Bryan E. Ogden

Department of Experimental Surgery, Singapore General Hospital

World Scientific

NEW JERSEY • LONDON • SINGAPORE • BEIJING • SHANGHAI • HONG KONG • TAIPEI • CHENNAI

Published by

World Scientific Publishing Co. Pte. Ltd.
5 Toh Tuck Link, Singapore 596224
USA office: 27 Warren Street, Suite 401-402, Hackensack, NJ 07601
UK office: 57 Shelton Street, Covent Garden, London WC2H 9HE

British Library Cataloguing-in-Publication Data
A catalogue record for this book is available from the British Library.

USING ANIMAL MODELS IN BIOMEDICAL RESEARCH

ISBN-13 978-981-270-663-8
ISBN-10 981-270-663-1

Printed in Singapore by Mainland Press Pte Ltd

ABOUT THIS BOOK

Biomedical research inevitably demands a well grounded understanding of the principles and practice involved with using *in-vivo* or animal models. The appropriate scientific use of such animal models come with proper training and an appropriate research environment, both of which do not happen spontaneously. The biomedical investigator has a duty both to science and to society to use animal models in a humane and rational way.

This book serves a few purposes. It is targeted firstly at the researcher who wishes to use animals for biomedical research and for this individual the book serves as a primer. The book also puts together between the covers of a single volume many necessary practical details required to appropriately manage a modern biomedical research animal facility, including the management of state-of-the-art bio-imaging equipment. To the established researcher finally, the book serves as a practical reference to the wide variety of uses animal models serve in biomedical research.

This book builds on a training manual first developed by the Department of Experimental Surgery, which together with the Institutional Care and Use Committee (IACUC) of the Singapore General Hospital started the first Responsible Care and Use of Animals for Scientific Research course on the Outram Campus in 8 May 2004. The course has been extremely popular and by June 2007 has trained more than 900 individuals from research institutions and universities (both local and foreign), the military and industry.

This course has benefited tremendously from the very beginning from the extensive knowledge of many individuals with expertise ranging from veterinary science, to statistics, radiation oncology and to the law. The underlying strength supporting all this has been the extensive experience of the Department of Experimental Surgery with using animal models in translational research, including the use of large animal models and non-human primates.

The final aim of biomedical research is better health for the community. We wish you every success in this endeavor.

FOREWORD

The crucial role of animals in advancing the medical research enterprise is indisputable, primarily ascribable to the close anatomic, physiological and genomic similarities of animals, especially non-human primates, and the *Homo sapiens* species. Hence, the utility of animal models for: (i) understanding the molecular basis of human disease processes, (ii) developing new therapies, and (iii) discovering drug toxicities prior to undertaking clinical investigation. Large animals are particularly useful for the pursuit of preclinical studies involving surgical and other interventional procedures, medical devices, drugs and prostheses.

The biomedical research landscape in Singapore is evolving rapidly, as highlighted by the government's recent announcement of the Biomedical Science Phase 2 initiative. A key aim of this laudable drive is to strengthen substantively our translational and clinical research capabilities along with the associated regulatory framework. Exemplary of the latter is the newly enacted legislature governing the care and use of animals for experimentation and teaching purposes. This progressive legislature requires institutions to obtain a license to conduct animal research and researchers themselves to receive formal training in handling research animals, thus helping to alleviate potential public concern regarding the proper conduct of animal research.

I am delighted to see that the Department of Experimental Surgery, Singapore General Hospital (SGH) and the SingHealth IACUC have been proactive in offering high-quality training courses to prepare individual investigators to launch their animal studies in a safe and technically sound manner. It is also refreshing to learn that the introductory course material is being updated and expanded in this *Primer*, which will undoubtedly serve as a valuable information source in the future.

It is both a pleasure and a privilege to write a foreword for this timely training manual. The contributors are experts in their respective areas and I thank them for giving generously of their knowledge and time. I wish all new researchers the very best in their future endeavors in animal investigation. May your studies prove to be both stimulating and rewarding, and bring credit to you, to your institution and, above all, to the nation as Singapore embarks on a well-funded program to conduct world-class translational and clinical research in order to advance the health and wealth of the island state.

Professor Malcolm Paterson
Chairman, Division of Clinical Research
SingHealth

PREFACE

The last decade has seen unprecedented advances made in the Biomedical and Life Sciences research. The completion of the sequencing of the Human Genome opens up infinite opportunities and potentials towards our search for answers to some of Medicine's most difficult problems. The proof of concept of these potential solutions would need to be carried out before trials can be contemplated in human subjects.

The Department of Experimental Surgery was set up in the early eighties with this as one of its main objectives. The use of animal models in these researches have proven to be an invaluable component of the whole chain of events leading up to eventual clinical applications.

Animal models are crucial test-beds for proof of concept prior to clinical trials. Appropriate choice of animal species and experimental models in these basic researches will help to provide directions for investigators towards their goals.

Over the years, the Department of Experimental Surgery has developed very stringent protocols and guidelines on the use of experimental animals, in line with accepted international standards of practice, ensuring that they are appropriately treated. Their safe handling, maintenance and welfare are no less stringent than what we would do for humans. In addition, high standards of bio-safety within facilities that carry out researches using such models must be observed in line with accepted best practices. The Department has been accredited by the Association for Assessment and Accreditation of Laboratory Animal Care International (AAALAC) in 2006 for these very high standards of practice, becoming the first institution in Singapore to achieve this. This benchmarks the Department internationally and is a reflection of the vision and commitment of the staff.

All investigators making use of animal models during the course of research work must ensure that they are appropriately trained and taught in the proper handling of these animals. Full compliance with guidelines by the host institution, regulating such use is absolutely essential to ensure successful completion of their projects.

Professor Ser Kiat Tan
CEO, SingHealth

CONTENTS

ABOUT THE EDITORS

ASSOCIATE PROFESSOR PIERCE K. H. CHOW

Pierce K. H. Chow (MBBS, Mmed (Surgery), FRCS(Edin), FAMS (Gen Surg), PhD) is Senior Consultant Surgeon at the Singapore General Hospital (SGH) and Associate Professor at the Duke-NUS Graduate School of Medicine. In addition to his busy clinical and teaching activities, he was also, from 2001 to 2007, the Director of one of the most established translational research institutions in Singapore, the Department of Experimental Surgery at SGH. He formed and chaired the first IACUC in Singapore, chaired the IACUC committee of NACLAR (National Advisory Committee on Laboratory Animal Research) and organized the initial training for IACUC members in Singapore. Under his charge, the Department of Experimental Surgery at SGH became the first research facility in Singapore to be awarded AAALAC accreditation. In his personal capacity as a researcher, he has taken novel medical innovations through conceptualization, product development and pre-clinical studies into clinical trials, thus demonstrating bench-to-bedside continuum and capabilities at SGH. His many awards include the Young Surgeon's Award of the Academy of Medicine (1995), the Seah Cheng Siang Prize (1997), the James Frazer Fellowship of the Royal College of Surgeons of Edinburgh (2002) and the Excellent Publication award of SGH (2006). He has more than 85 publications on MEDLINE. He serves on the MedTech Concept Advisory Committee of the Economic Development Board.

MR. ROBERT T. H. NG

Robert T. H. Ng (Dip Med Tech) is Senior Manager at the Department of Experimental Surgery at SGH and has a vast and practical experience of more than 20 years, with animal models in biomedical research. In his professional work he has helped create and develop a wide variety of animal models from rodents to non-human primates and co-authored many of the publications. He has extensive experience serving on the IACUC in SingHealth.

DR. BRYAN E. OGDEN

Bryan E. Ogden (DVM) is the Institutional Veterinarian at SGH and at Maccine Pte Ltd. As a Diplomate in the American College of Laboratory Animal Medicine he is a recognized expert in Laboratory Animal Science. He has extensive experience in animal models including the use of nonhuman primates. Included in his long and distinguished career are time spent as Attending Veterinarian at the Oregon National Primate Research Center and the Oregon Health & Science University. He was instrumental in the formation of the Singapore Association for Laboratory Animal Science (SALAS), currently serves on the IACUCs at SingHealth and Maccine Pte Ltd, is a member of Singapore's National Advisory Committee for Laboratory Animal Research (NACLAR), and is an *ad hoc* site visitor for the Association for Assessment and Accreditation of Laboratory Animal Care International (AAALAC).

LIST OF CONTRIBUTORS

Chow Pierce K. H., *M.B.,B.S, FRCS (Edin), M.Med (Surg), FAMS (Gen Surg), PhD*
Senior Consultant Surgeon, Dept of General Surgery, Singapore General Hospital
Associate Professor, Duke-NUS Graduate Medical School

Goh Angela, *B.Sc (Hons)*
Former Research Associate, Dept of Experimental Surgery, Singapore General Hospital

Kee Irene, *B.Sc (Hons)*
Senior Research Associate, Dept of Experimental Surgery, Singapore General Hospital

Kuah Boon Theng, *L.LB (Hons), M.A (London)*
Advocate and Solicitor, Legal Clinic LLC
Member, SingHealth-IACUC

Li Huihua, *Ph.D*
Biostatistician, Clinical Trials & Epidemiological Sciences, National Cancer Centre
Member, SingHealth-IACUC

Lin Manjing, *Ph.D*
Research Scientist, Dept of Experimental Surgery, Singapore General Hospital

Ng David, *M.B.,B.S, MRCP, FAMS*
Senior Consultant, Dept of Nuclear Medicine & PET, Singapore General Hospital
Member, SingHealth-IACUC

Ng Robert, *Dip Med Tech*
Senior Manager, Dept of Experimental Surgery, Singapore General Hospital
Member, SingHealth-IACUC

Ogden Bryan E., *DVM, DACLAM*
Institutional Veterinarian
Member, SingHealth-IACUC

Ong Hock Soo, *M.B.,B.S (S'pore), FRCS (Edin), FAMS (Gen Surg)*
Senior Consultant, Dept of General Surgery, Singapore General Hospital
Member, SingHealth-IACUC

Ong Lai Chun, *B.Sc (Hons)*
Research Associate, Dept of Experimental Surgery, Singapore General Hospital

Phua Cindy, *B.Sc (Hons)*
Research Associate, Dept of Experimental Surgery, Singapore General Hospital

S. Somanesan, *B.Sc (Hons)*
Senior Principal Radiation Physicist, Dept of Nuclear Medicine & PET, Singapore General Hospital
Member, SingHealth-IACUC

Sergio Darvi, *DVM*
Assistant Veterinarian, Dept of Experimental Surgery, Singapore General Hospital

Song In Chin, *Dip Med Tech*
Senior Research Associate, Dept of Experimental Surgery, Singapore General Hospital

Tan Bien Keem, *M.B.,B.S (S'pore), FRCS (Edin), FAMS (Plast. Surg)*
Senior Consultant, Dept of Plastic Reconstructive & Aesthetic Surgery, Singapore General Hospital

Tan Bien Soo, *M.B.,B.S (S'pore), FRCR*
Head and Senior Consultant, Dept of Diagnostic Radiology, Singapore General Hospital

Tan Peik Khin, *M.Sc*
Assistant Veterinarian, Dept of Experimental Surgery, Singapore General Hospital

Tham Colin, *M.B., ChB. (Aberdeen), FRCS (Ireland)*
Consultant, Dept of Plastic, Reconstructive & Aesthetic Surgery, KK Women's and Children Hospital

Then Inria Kurniawan, *M.Sc*
Executive Research Coordinator, Dept of Experimental Surgery, Singapore General Hospital
Member, SingHealth-IACUC

Thng Choon Hua, *M.B.,B.S, FRCR, FAMS*
Senior Consultant Radiologist
National Cancer Centre, Singapore

Villano Jason, *DVM M.Sc*
Assistant Veterinarian, Dept of Experimental Surgery, Singapore General Hospital
Member, SingHealth-IACUC

Woo Evan, *M.B.,B.S (S'pore), MRCS (Edin), M.Med (Surg)*
Registrar, Dept of Plastic Reconstructive & Aesthetic Surgery, Singapore General Hospital

Yong Peggy, *B.Sc (Hons)*
Research Associate, Dept of Experimental Surgery, Singapore General Hospital

Yu Sidney, *Ph.D*
Senior Principal Research Scientist, Dept of Nuclear Medicine & PET, Singapore General Hospital

Zhang Kai, *M.B.,B.S, Ph.D*
Senior Scientist, Dept of Experimental Surgery, Singapore General Hospital

Zheng Lin, *M.Sc*
Research Associate, Dept of Experimental Surgery, Singapore General Hospital

CHAPTER 1

SCIENTIFIC CONSIDERATIONS AND CHOICE OF SPECIES

CHAPTER **1.1**

THE RATIONALE FOR THE USE OF ANIMAL MODELS IN BIOMEDICAL RESEARCH

Pierce Chow

The use of animal models in biomedical research, is accepted by the majority of lay people and scientists alike as being necessary to the advancement of useful knowledge that brings about relief from suffering. Few outside of the biomedical scientific community, however, have a clear understanding of why these animal models are important. This is unfortunate. Animals and man are symbiotic in many real ways and not just on an ideological level. Arguments regarding whether biomedical science can advance without the use of animals are frequently mooted and makes as much sense as questioning if clinical trials are necessary before new medical therapies are allowed to be widely used in the general population.

Addressing these questions has, however, become increasingly urgent with the spectre of both bio-terrorism and the increasing use of therapies derived from biological systems. While the use of animals has apparently declined in the last two decades, advances in genetic research and the demands of research to counter bio-terrorism are expected to reverse this trend and lead to an increase in animal use. At the heart of it all is the health and safety of human populations.

The rationale for using animal models in biomedical research is scientific and animal models are likely to remain necessary until science develops alternative models and systems that are equally sound and robust. This chapter discusses the nature of experimental research, the role of models in science and the relevance of these in biomedical research.

The Place of Experimental Studies in Biomedical Research

Scientific research implies the systematic and empirical investigation of hypotheses. In the biomedical sciences such systemic investigations may be classified as observational or experimental.

Observational studies are frequently (and most usefully) carried out when the variables influencing the outcomes of the phenomena under study either cannot be controlled directly or cannot be easily manipulated. These variables are, thus, carefully observed (occasionally over long periods of time) and an attempt is made to explain or determine the correlations between them. Examples include observing animals in their natural habitat and understanding how recent ecological changes impact on their survival.

Such observational studies also abound in clinical medicine and include *descriptive case series, retrospective (case control) studies, prospective (cohort) studies and cross-sectional studies* or surveys. These studies are particularly important where the conditions are rare and where it is important to understand the natural history of a particular condition (including the outcomes of currently accepted therapy). Investigations of these nature are also common in the basic biomedical sciences e.g. in molecular epidemiology and in comparative anatomy.

Experimental studies require intervention and attempts are made to directly control selected variables and to measure the effects of these variables on outcome. Such studies are necessary to establish cause and effect relationships in an unequivocal and rigorous manner. The results of experimental studies tend to be more robust compared to observational studies (although not necessarily more important) and many breakthroughs in the biomedical sciences are made possible only through experimental studies. Data arising from interventional studies also lend themselves easily to statistical analysis.

The definitive experimental study in clinical medicine is the *randomized controlled trial* some of which can run into large numbers of patients. Clinical trials could of course also be carried out in non-randomized manners such as in *sequential self controlled* or *cross over trial* protocols. There are legislative requirements for most if not all new therapies (such as pharmaceutical or device related) to have undergone rigorous clinical trials before being accepted in mainstream medical practice and be considered standards of care.

Experimental studies are similarly important in preclinical biomedical research. These studies may be carried out on ***in vitro biological systems*** such as isolated cells, cell culture systems, tissue slice preparations or isolated perfused organs. Experiments using *in vitro* systems are particularly useful in the early phases of studies where the screening of large number of potential therapeutic candidates may be necessary.

In vitro systems are, however, by definition, nonphysiological and have important limitations. Living creatures are biologically complex and this especially true in higher order animals including man. While data from experiments carried out in *in vitro* systems can establish mechanisms and define toxicities, ***in vivo biological systems*** using live animals (whole organisms) are necessary to study how such mechanisms behave under clinical or pathophysiological conditions.

Intact (whole) animal systems are, thus, extremely important for "proof of principle" research. It is frequently possible to have a clearer understanding of the efficacy, pathophysiological interactions and potential toxicities of novel therapies only with whole *in vivo* biological systems. Many *in vivo* interactions are complex and cannot be predicted from *in vitro* data. Such information is especially important when assessing the safety and efficacy of biologics. Biologics are therapeutics (drugs, vaccines, antibodies etc.) synthesized from living organisms. Biologics have made great advances in the last decade through advances in genetics and molecular biology especially recombinant DNA technology. Such therapies are increasingly developed and have contributed significantly to better outcomes in diseases e.g. in cancer therapy.

Models in Biomedical Research

A model is "a representation of a real or actual object" (Oxford English Dictionary). Models are, thus, meant to mimic and it is not expected that a model be necessarily identical to the subject under study. Models are widely used in all branches of physical, biological and social sciences. In biomedical research, models allow the investigator to understand and investigate pathophysiological processes and the impact of intervention. As described above, these models can be *in vitro* or *in vivo*.

Biomedical research models can also be either analogues or homologues. ***Analogous models*** relate one structure or process to another and are not unique to biomedical research. Such models are also common in physics, engineering and mathematics. A scaled-down model of an aeroplane is not an aeroplane but allows appreciation of how the various parts of the structure relate to one another and how improvements may be usefully made. Similarly, large animal models like the pig allow the development of new minimally invasive surgical techniques and instruments.

Homologous models reflect counterpart genetic sequences and are only used in biomedical research. Many animal models are both analogues and true homologues.

The ideal model for a human is another human, which is why randomized controlled clinical trials will always be important in the evaluation of new therapies. Famous historical examples using a human subject as a model will of course include Edward Jenner's classical "proof-of-principle" experiment of the efficacy of inoculation against smallpox using a hapless farm boy as a subject, presumably without informed consent (and without the approval of an Institutional Review Board!).

Research using human subjects is only justified and should only be allowed if there is sufficient understanding of the underlying mechanisms of action and of the bio-safety parameters involved in the research. Robust preclinical data of this nature are most accurately derived from the use of animal models and must pass the scrutiny of institutional review boards and health authorities. This is especially important with "first-in-man" studies of novel therapies. The use of higher order animal models with close genetic homology to man, such as nonhuman primates, is particularly important in studies involving therapeutics derived from biological systems i.e. biologics.

Animal Models in Biomedical Research

In biomedical research, an animal model is defined as "a living organism with an inherited, naturally acquired or induced pathological process that in one way or another closely resembles the same phenomenon in man" (Wessler 1976). The ultimate goal of experimental research using animal models is to solve problems in clinical practice and to develop new methods and approaches to the cure and alleviation of disease and disability (Isselhard, Kushe 1986).

Both invertebrate and vertebrate animals are used as models in biomedical research. ***Invertebrate models*** are very useful in the fields of neurobiology, genetics and development and notable examples of invertebrates use for such purposes include the *C. elegans* and *Drosophila*.

Vertebrate models are responsible for many advances in biology and medicine and are extremely important in translational research. This includes the use of both small animal models (e.g. mice, rats, and rabbits) and large animal models (e.g. dogs, pigs, and monkeys).

Broad areas of how vertebrate animal model are used in biomedical research include:

1. *Pharmaceutical research including the development of biologics*

2. *Toxicology testing*

3. *Development and testing of new medical devices*

4. *Surgical research*
   a. the development of new surgical techniques e.g. techniques of gastrectomy, open heart surgery, coronary artery surgery, microsurgery, endoscopy and the use of arterial ligation in treating aneurysms (by the pioneer surgical scientist John Hunter).
   b. the development of new therapies e.g. organ and tissue transplantations, cardio-pulmonary resuscitation.

5. *Pathophysiological research*
   Animal models were crucial to the understanding of basic and important patho-physiology processes such as shock and the body's response to trauma, regeneration and malignancy. In particular the development of the concept of the "milieur interieur" in physiology (by the pioneer physiologist Claude Bernard) and the concept renal dialysis all depended on the use of animal models.

The above is not exhaustive. The vast majority of animals used in biomedical research are in the fields of pharmaceutical research and toxicology testing.

When animal models are used for therapeutic testing, an established principle is to use the minimum number of animals necessary to arrive at scientifically robust data and to ensure the humane and proper care of animals so that the scientific data is reliable. Generally, two or more species (one rodent, one non-rodent) are tested because a drug may affect one species differently from another. Besides treatment efficacy, animal models are also used to determine how much of a drug is absorbed into the blood, how it is broken down chemically in the body, the toxicity of the drug and its breakdown products (metabolites), and how quickly the drug and its metabolites are excreted from the body.

What Makes a Good Animal Model?

Not all animal species are useful for the purposes of biomedical research and the limitations of the models selected as well as the methodology involved must always be kept in mind. Biomedical research is a very vast field and there are both general and specific uses for animal models. In the early years of biomedical research, animal models were mainly used for general research purposes, that is to uncover broad pathophysiological phenomena and principles. The recent development and widespread use of transgenic animal models in biomedical research have made many animal models very specific to the nature of individual research projects.

While there are always exceptions, a good and useful animal model suitable for ***general use*** in a research facility should have the following characteristics (adapted from Isselhard, Kushe 1986):

1. The animal model should closely reproduce the disease or condition under study.

2. The animal model should be easily available to many researchers, that is, not a rare or exclusive animal. This allows validation and stimulates further investigations.

3. The animal model, in the case of a vertebrate model should be large enough for multiple biological sampling (e.g. tissue and blood).

4. The animal model should fit into available animal facilities of the average institution.

5. The animal model should be easily handled by most investigators.

6. The animal model should be available in multiple sub-species.

7. The animal model should survive long enough for results to be meaningful.

8. The animal model should be sufficiently robust for the purpose of the study.

Transgenic animal models, spontaneous animal models (see below) and highly specialized animal models such as nonhuman primates do not fit these traditional guidelines. Such ***special animal models*** are, however, increasingly used in biomedical research

Consideration in the Selection of an Appropriate Animal Model

The researcher should consider using established models where possible or available (Table 1.1.1). The model must, however, be relevant to the aims of the study. The following serves as examples:

1. *Relevance of species*
   For example, animals are suitable for studies on muscle contraction but data obtained from the whole body has little relevance to humans. In gastrointestinal tract and liver

studies, herbivores have highly specialized gastrointestinal parts (e.g. for cellulose digestion) and associated metabolism, which has no counterparts in humans. Omnivores are, thus, most suitable e.g. pigs.

2. *Numbers required*
   In studies where the outcomes between the control and study groups differ only in degree, large numbers of animals are required to achieve statistical significance. Mice and other small mammals are ideal.

3. *Transplant and other immunological studies*
   Inbred or naturally immunosuppressed species may be required.

The animal model that is required to address the specific research question may, however, not have been previously developed or validated in some instances. The research effort must then begin by developing and validating a suitable model rather than using an established one. The development of a suitable model in this case becomes critical because it is essential that the model be reliable, reproducible and valid. The model must also be a reasonable representation of the actual situation and the limitations of the model must be identified. The validity of the results in experimental research depends on the qualities of the experimental model.

Table 1.1.1: Examples of established general animal models

Models	Species
Haemorrhagic Shock	Rat, rabbit and pig
Stress Ulcers	Rat restrain model
Hypercholesterolaemia	Minipig
Sepsis Model	Rat, dog and pig
Primary Liver Cancer	Rat
Liver Regeneration	Rat and pig
Acute Pancreatitis	Dog and rat
Inflammatory Bowel Disease	Rabbit
Myocardiac Infarction	Baboons
Vascular Grafts	Dog, pig and sheep
Bone Fracture	Rabbit

Specific Animal Models

Occasionally, researchers may seek to use animal models that specifically mimic conditions of interest as opposed to using or developing general models. Such animal models may either spontaneously mimic these conditions or be induced to simulate those conditions.

Spontaneous animal models are those models that have arisen through spontaneous mutations to mimic specific conditions. Notable examples of these are the Gunn rat (for hereditary hyperbilirubinemia) and the BB Wistar rats (for type I diabetes).

Induced animal models can be created through surgical manipulations, chemical manipulation and genetic manipulations (including negative models).

The surgically induced model is in many ways the classical biomedical research model and was used to understand brain plasticity (nonhuman primates), develop organ transplantation (dogs and pigs), discover the role of insulin in diabetes (dogs) and to develop cardio-pulmonary resuscitation (dogs).

Examples of chemically induced models include the chemical ablation of beta cells to create diabetes (rats, rabbits, pigs, monkeys) and the use of carbon tetrachloride to create cirrhosis (rats).

Transgenic animal models are important induced animal models. A transgenic animal is one that carries a foreign gene that has been deliberately inserted into its genome. An example of a transgenic animal model is mice with type I diabetes ($Cd38^{tm1Lnd}$). Homozygous mutant mice show impairment in glucose-induced increases in ADP-ribosylcyclase/cyclic ADP-ribose (cADPR), intracellular calcium concentrations and insulin secretion.

Some Special Roles of Animal Models

In the development of drugs against bio-terror agents, controlled studies of clinical effectiveness in humans are unethical. Since generally few people would have been previously exposed to these agents/diseases and have been treated, observational studies may not provide sufficient data.

Under these circumstances the role of animal models becomes especially important. Instead of depending on human studies, the FDA allows approval of drugs shown to be effective in two animal models, without clinical trials for effectiveness. Examples of such circumstances are treatment against anthrax, botulism, plague, smallpox, tularaemia and viral haemorrhagic fevers.

A Short History of the Use of Animal Models in Biomedical Science

In the western scientific tradition, the initial use of animal models was in experimental surgery, which pre-dated all other scientific uses by more than a millennium. In antiquity, the earliest records of physiology research were carried out by Erasistratus of Alexandra (302 – 258 BC) on the functions of the heart and respiratory systems in pigs. The first textbooks on anatomy by Galen (129 – 200) were based on dissections not on human cadavers (which was forbidden by religious and legal authorities) but on pigs and apes. Although these observations and their interpretations were frequently erroneous, they established the discipline of comparative anatomy. While animal models remain central to the development of new surgical techniques and the invention of novel medical devices, the number of animals used in experimental surgery today is only a small fraction of the total number of animals used in biomedical research.

In 1628, William Harvey published his great work on circulation based on studies in animals. The "father" of modern physiology, Claude Bernard (1813 – 1895) established the basis of the discipline based on animal experimentation and Louis Pasteur (1822 – 1895) used animals in the validation of the experimental method in microbiology. In the 20th century, cardio-pulmonary resuscitation, the discipline of immunology and translational research on organ transplantation were all primarily developed through the use of animal models. Koch's postulate for the carcinogenesis of the *Helicobacter* bacteria was fulfilled in gerbils in the 1990s.

The explosion in molecular biology in the second half of the 20th century increased the importance of *in vivo* models. In the 1980s, the pathology of Hepatitis C was established through infecting chimpanzees with the virus. Examples of other diseases where the use of animal models were crucial to the recent elucidation of pathogenesis include cystic fibrosis, rheumatoid arthritis and spongioform encephalopathies. The use of naturally immunosuppressed animals such as SCID and nude mice to harbour cancer cells were similarly crucial to the development of experimental oncology and new therapies in cancer. Increasingly, animal models are now being produced to exhibit specific symptoms and pathology of diseases through selective breeding and genetic modification.

The development of ***in vivo molecular imaging*** modalities such as the micro-PET and MRI and their application to animal models in the 21st century has brought about a degree of accuracy and sophistication on biomedical research not previously possible. Such *in vivo* imaging and documentation of cellular processes in animal models confers increased scientific vigour to experimental design and leads to fewer animals being required in each experimental protocol. The robustness of such data increasingly contributes to the ease of translation of biomedical breakthroughs from preclinical studies to clinical applications.

The Limitations of Animal Models

All models have their limitations concerning transferability and predictability and this is true in every branch of science. The extent of the validity of extrapolating data derived from specific experiments using animal models to the general human clinical conditions depends on the degree to which the animal model is an appropriate reflection of the condition under investigation, the design of the experiment and the technical experience of the researchers.

These limitations are, however, an intrinsic part of all modelling approaches that use surrogates and do not render the scientific method invalid. They are also similarly found in clinical and *in vitro* studies. This explains why unexpected adverse reactions can sometimes still occur when medicines are brought into the market even after extensive clinical trials.

The question of the scientific validity of data derived from animal models is often confused with questions pertaining to complex ethical issues. The separation of science and ethics is important in such discussion. Each scientific study has to be judged on its own merits after careful evaluation of the methodological and statistical rigor. Scientific experience and balance are important attributes in such judgement.

A recent review by the Nuffield Council on Bioethics concluded that "animal research has been, and can potentially be, scientifically valid, in that it is possible to extrapolate from animal models to humans (or other animals)...." (Nuffield 2005). The Council further cautioned that data on the validity of animal experiments have been interpreted and used in different ways by both opponents and proponents of the scientific validity of using animal models.

The public health perspective on the use of animal models in scientific research is, however, unequivocal. It is unlikely that any health authority will allow novel therapies in medicine be approved for use in the general population without scientifically rigorous supporting animal and clinical data. Likewise, responsible Institutional Review Boards are unlikely to allow clinical trials on novel therapies to be carried out within that institution without supporting animal data.

Improvement in the technology used in animal research (such as *in vivo* molecular imaging) continually refines the interpretation of data derived from animal models today. Together with improvement in methodology (e.g. the use of orthotopic models and tumour explants in experimental oncology), there is an expectation that the extrapolation of data derived from animal models to the human condition will be even more valid in time to come.

CHAPTER **1.2**

EXPERIMENTAL ANIMAL MODELS IN BIOMEDICAL RESEARCH

Robert Ng

The use of animals in *in vivo* experiments is an essential feature of any Experimental Surgery Laboratory. In the choice of species, the most important considerations are the goals of the study, the constraints imposed by the animal study design, the familiarity of the laboratory with a particular species, cost and relevance of the obtained data for clinical use.

Mice and Rats

Small animals such as rats and mice are preferred for many experimental studies as they have short life cycles, are inexpensive to purchase and easy to maintain in limited space. They are extensively used in veterinary and biomedical research for the discovery and manufacture of vaccines, antibodies and hormones and in the testing of drug potency and toxicology. In pharmaceutical industries, rodents are widely used to study the metabolism of drugs and food additives, since extrapolation of their results to humans is a necessary prerequisite to clinical trials. In addition, when a large number of graded doses of an investigational therapeutic agent are being examined and when larger number of animals is needed for studies with survival as an endpoint, economic feasibility takes priority as a consideration. The following information is mostly specific to studies conducted at the Department of Experimental Surgery, Singapore General Hospital.

A. *Rats*
Rats have been used in experimental neurosurgery and in cardiac transplantation. Research on abdominal heart allografting is also a well established model in the rat.

And, the rat body size is suitable for microCT imaging. Some of the rat models used include:

- **Outbred Wistar rat**
  This outbred rat has become a standard feature in most experimental laboratories today and has its origin in the Wistar Institute in Philadelphia, and was name after Professor Casper Wistar (1761–1818). Good breeders have long lifespan (30 months) and are excellent parents. However, they have the tendency to become very fat with age.

- **Outbred Sprague-Dawley rat**
  This strain was developed in 1925 by Robert W. Dawley, who named the strain after himself and his first wife, whose maiden name was Sprague. A commercial firm, Sprague-Dawley Inc., was subsequently set up in Wisconsin and was dedicated to the establishment and sale of this rat strain. Descendants of the Sprague-Dawley strain today are random bred and extremely popular in experimental laboratories worldwide. These rats are highly intelligent and are thus popular models for psychology tests.

- **Inbred Buffalo and Fisher 344 rats**
  At the Columbia University in 1921, five inbred lines were initiated using rats from four local vendors namely August, Fisher, Marshall and Zimmerman and one vendor from Copenhagen. From this first litter of pedigree, rats after the 344th mating was subsequently derived as the well-known Fisher 344 strain. The Buffalo and Fisher 344 were two of the seven inbred strains that were established in 1953 at the National Institute of Health and these inbred were used as models for oncogenesis. Cell line Morris 7777 is specific for the development of hepatocarcinoma in Buffalo rats (Fig 1.2.1).

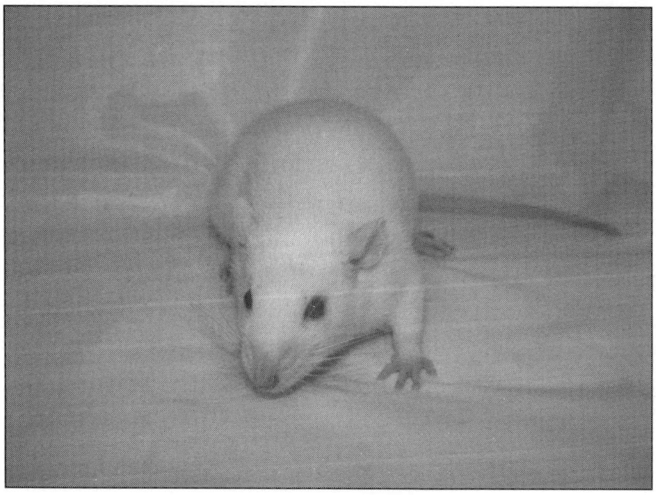

Fig 1.2.1: Inbred Buffalo rat.

- **Inbred Brown Norway rat**
  This rat has moderate incidence of urinary bladder and ureteral carcinoma (35%) in old males. There is also moderate incidence (14% male, 26% female) of pituitary adenomas in older animals. This rat is a good immunologic model as the strain is unique in its expression of high IgE-responder phenotype. They are resistant to experimental allergic encephalomyelitis and to induction of autologous immune complex glomerulonephritis. Poor to moderate breeders with moderate lifespan of approximately 25 months for male and 28 months for female.

- **Inbred Dark Agouti rat**
  Its uses include studies on transplantable salivary gland adenocarcinomas, transplantation immunology and autoimmune disease, for example thyroiditis.

- **Inbred Lewis rat**
  The rat is susceptible to induction of experimental allergic encephalomyelitis, adjuvant induced arthiritis, induced autoimmune myocarditis and autologous immune complex glomerulonephritis. It is also a host for a number of induced neoplasms (lymphoma 8, renal sarcoma and fibrosarcoma).

- **Mutant Athymic Nude rat**
  These are homozygotes having little or no hair and lack of thymus gland (referred to as athymic). They have similar dysgenesis to the nude mutation in mice. Cell mediated immunity is greatly reduced or absent with marked reduction of T-lymphocyte function. They are thus suitable for tumour xenograft studies. They are extremely susceptible to infection with *Clostridium piliformi.*

B. Mice

Mice are mainly used for experimental oncology research and their small sizes are most suitable for bioimaging procedures like microCT and microPET scanning. The most commonly used mice include the following:

- **Outbred Swiss mice**
  This is a general-purpose mouse recommended for dissection and any work not requiring the special qualities of inbred strains.

- **Inbred BALB/c mice**
  This mouse is susceptible to chronic pneumonia and extremely sensitive to radiation. BALB/c (Fig 1.2.2) has a high level of alpha-foetoprotein. Commonly used for ascitis fluid production. Injection of mineral oil intraperitoneally induces a high incidence of transplantable plasmacytomas. They have low mortality after neonatal thymectomy.

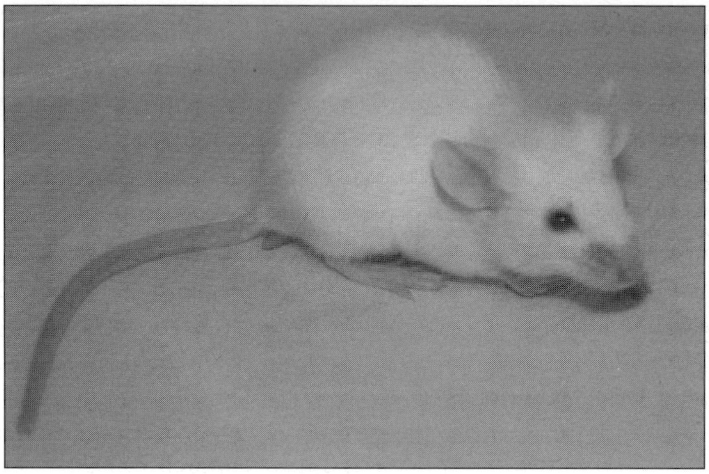

Fig 1.2.2: BALB/c mouse.

- **Inbred C57BL mice**
  This is an intense black mouse with very low incidence of mammary tumours. Used most commonly as the background for various mutant genes (for example, nude, beige and knockouts). These mice (Fig 1.2.3) are used in alcohol tolerance studies and are susceptible to the development of atheromatous lesions after 20 weeks on high fat diet. When fed high fat, high simple carbohydrate diet, they develop noninsulin dependent diabetes mellitus and hypertension. They are notorious for developing a progressive, ulcerative dermatitis, which may be responsive to an altered fatty acid diet.

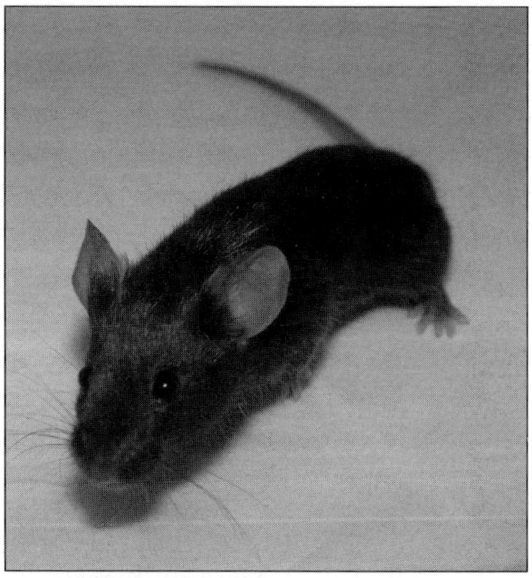

Fig 1.2.3: C57BL mouse.

- **Inbred CBA/CaH mice**
  This mouse line is known to be missing the lower third molar in about 18% of the offspring. They do not develop antinuclear antibodies or LE cells with ageing. They are resistance to *S. typhimurine* and *Leishmania* infection, but are highly susceptible to the Edmonton strains of measles virus.

- **Mutant BALB/c Athymic nu/nu mice**
  Nude is an autosomal recessive mutation located on Chromosome 11. The two major defects are failure of hair growth and dysgenesis of thymic epithelium. Nude mice (Fig 1.2.4) respond poorly to thymus-dependent antigens because of a defect in helper T-cell activity. They have often been used for tumour xenotransplantation studies.

Fig 1.2.4: Mutant Balb/c Athymic nu/nu mouse.

- **Mutant SCID mice**
  The severe combined immunodeficient (SCID) mouse arose as spontaneous autosomal mutation in C.B-IgH-I (CB17) congenic strain. Homozygotes have little or no immunoglobin in serum. Lymphoid organs consist of vascular connective tissue and macrophages and are devoid of lymphocytes. Although B and T-cells are absent early B and T-cells are present. They are a useful model for studying the relationship between immunity and disease, studies on engraftment of xenogenic cells and tissues and studying human severe combined immunodeficiency.

Pig

Apart from primates, the pig (Fig 1.2.5) is thought by many to be the laboratory animal species nearest to humans in terms of anatomy and physiology. Being relatively inexpensive to obtain and maintain, the pig is quite popular for physiological and pharmacological studies. However, the rapid growth rate of the domestic pig makes it unsuitable for chronic experiments. For experiments where there is need for post operative maintenance of six months to 2 years, the mini- or micro-pig (Gottingen, Yucatan, PWG or Bama) should be used, Micro-pig's weight at two years is less than 50 to 60 kg as opposed to the normal domestic pigs like Yorkshire or Landrace where weight can go beyond 200 kg for the same period. In spite of this limitation, the pig is still the animal of choice for colorectal, angioplasty and transplantation surgeries or other acute procedures as its size and anatomy mimic the human structure. It is also suitable for experiments on trauma management as in dermal burns, resuscitative shock or traumatic brain injury.

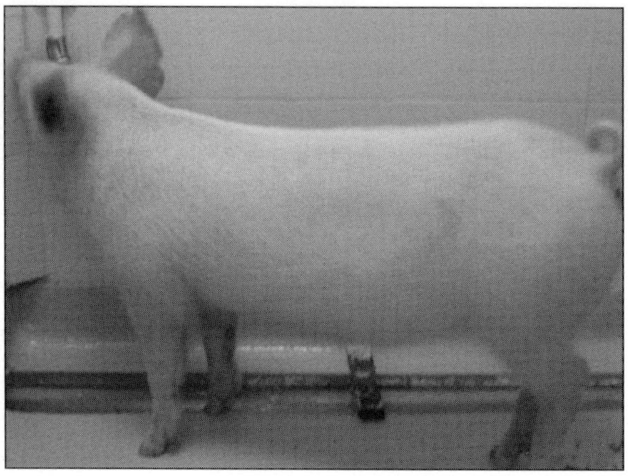

Fig 1.2.5: Pig.

Sheep

Domestic sheeps are placid animals of manageable size and are choice animals in several areas of biomedical research. They tolerate implanted electrodes and indwelling catheters in blood vessels and lymphatics better than most other species and make little effort to remove them. Hence, they are often used for chronic experimental preparation for studies of endocrine function and for immunological or isotope research involving chronic collection of lymph draining from different anatomical areas. They also recover well from foetal instrumentation surgeries. Sheep is an appropriate animal for cardiothoracic surgeries such as heart valve implantation in which they have shown high resilience against vascular insult during such surgeries and a lower tendency towards thrombosis formation than many species.

Rabbit

New Zealand White is a popular non-inbred strain for various research projects especially for polyclonal antibody production. It is a commonly used animal model for research studies involving orthopaedic surgery and ophthalmology. Rabbit (Fig 1.2.6) is also widely used for paediatric intensive care courses as it offers good simulation of human infant patient for chest tube placement.

Fig 1.2.6: Rabbit.

Golden or Syrian Hamster

This animal is agouti or golden yellow in colour with many coat mutations. An excess vitamin in the diet results in fatty liver and teratogenesis. It is used for a variety of experimental purposes including immunology, parasitology, reproductive physiology and cancer research.

Guinea Pig

The white strain, Hartley is the most commonly used outbred stock for routine and experimental procedures such as skin patch test for efficacy test on antiseptics and antibiotics efficacy test. This animal has a tendency to be more prone to deafness than non-white stock.

Nonhuman Primates

Macaca fascicularis (Fig 1.2.7) is the common nonhuman primate used for experimental research in Southeast Asia. Although costly to work with, they remain invaluable for selected studies by virtue of their similarity to humans. However, nonhuman primates are generally reserved for preclinical experiments when only small sample size is needed.

Fig 1.2.7: ***Macaca fascicularis*** **(longtail or cynomolgus macaque).**

CHAPTER **1.3**

NONHUMAN PRIMATES AS MODELS IN BIOMEDICAL RESEARCH

Jason Villano and Bryan Ogden

Nonhuman primates (NHP) belong to the order Primates, which contains two suborders:

1. **Strepsorrhini**, the 'wet-nosed primates', which include the lemur and the loris; and

2. **Harplorrihini**, the 'dry-nosed primates' which are the true primates divided into infraorders, Tarsiformes (tarsiers) and Simiiformes. Simiiformes is further divided into parvorders:

   a. Platyrrhine or New World monkeys (NWM) – found in Central and South America. Examples include marmoset, tamarins, squirrel monkeys and capuchins.
   b. Catarrhine or Old World monkeys (OWM) and hominids, including man, chimpanzees, and gibbons.

The Old World monkeys will be the focus of this discussion.

Old World Monkeys (OWM)

The Old World monkeys (OWM) of Africa and Asia belong to the family Cercopithecidae. Their common characteristics include having a specialized digestive mechanism for processing a folivorous diet, narrow noses with comma-shaped nostrils separated by a narrow nasal septum, ischial callosities, opposable thumbs, marked sexual dimorphism, and cheek pouches.

To this group belong the most commonly used NHPs in biomedical research: the *Macaca fascicularis*, also known as cynomolgus or crab-eating or longtail macaque and the *Macaca mulatta*, the rhesus macaque.

- *Macaca fascicularis* (Fig 1.3.1) — found mainly throughout Southeast Asia, including Singapore. Its tail is longer than its head and body, which is brown with grey or black tones with crown hair, directed either backward and outward or in a crest. Adult females generally weigh from 2 to 6 kg and adult males from 4 to 8 kg. But, there are ten sub-species of *M. fascicularis* each with distinctive fur, morphometric and genetic differences.

- *Macaca mulatta* — found in the northern half of the Indian continent, Northern Burma and Indochina, and much of China. It has a medium length tail and is brown with a reddish tone on its hind parts, including hindlegs. Adult females weigh 4 to 9 kg and adult males weigh 6 to 11 kg, though weights exceeding this range are not uncommon. As with *M. fascicularis*, there are regional differences within the species, most notably immunologic differences between Indian- and Chinese-rhesus as noted by AIDS researchers.

In Singapore's booming biomedical research industry, the cynomolgus macaque is the most widely used NHP since it can be found locally and on some nearby islands like Bintan, Indonesia.

Anatomy and Physiology

The phylogenetic proximity between NHPs and humans is reflected by many similarities between them. From an anatomical perspective, NHPs are the closest model to humans among many laboratory animals. Similarities include, but are not limited to, skeletal structure, growth and development, organ structure and dentition. Physiological similarities include immune function, neuroendocrine function and processes such as aging of the brain and other body systems. These and other similarities make them a preferred model for biomedical research in such areas as pharmacology (especially with biologics), immunology, pathobiology, neurobiology, behavioural neuroscience and aging processes.

Current Uses in Biomedical Research

The use of nonhuman primates in research has led to significant discoveries during the past 100 years. Studies on relapsing fever, typhoid and yellow fever benefited from NHP use. The discovery of Rh factor depended on work in rhesus macaques. The role of NHPs has expanded to cover a multitude of research areas, including pharmacology, aging, metabolic disorders, gene therapy (Fig 1.3.2), virology and many others. A summary of a few of these areas follows:

1. *Pharmacological research*
   Animals have been essential in pharmacokinetics and toxicological studies. Besides establishing the efficacy of a certain compound to produce a desired effect and the specifics of its ADME (absorption, distribution, metabolism, and excretion), the safety of a compound must be established in a rodent and a non-rodent species (usually dogs, but increasingly NHPs). While research involving nonhuman primates provides a meaningful translation towards understanding human disease and the development of treatments, their similarities to humans also becomes the drawback to their use as significant bioethical issues must be addressed. Many of these issues are discussed elsewhere in this text. Justifications for using NHPs for pharmacological and toxicological research often depend on the characteristic of the test compounds. For example, large molecules or biologics need to be tested on species with the most immunologic similarity to humans. Also, the receptors targeted by some molecules are unique to human and nonhuman primates.

2. *Human aging and metabolic disorders research*
   There has been substantial interest regarding primate aging. Study areas involving NHPs included the neurobiology of aging as well as reproductive senescence. Metabolic disorders such as obesity and diabetes, spontaneously-occurring and experimentally-induced, have been frequently studied in NHPs. Macaques will consume carefully formulated high fat, high cholesterol diets. This and their similarity to humans make them susceptible to many of the other chronic diseases, such as atherosclerosis, that affects human populations.

   Streptozotocin is a naturally occurring chemical that is particularly toxic to the insulin-producing beta cells of the pancreas in mammals. While it is used as medicine for treating certain cancers of the islets of Langerhans, it is also used in biomedical research to produce Type II diabetes in NHPs and other laboratory animals.

3. *Neurological research*
   Nonhuman primates offer a valuable choice of animal model in studies of neurological disease and cognition. With their cerebral organization approximating that of humans, they can be used to study cognition, functional connectivity, and neurotransmitter pathways. It is important to note that certain disorders in humans occur spontaneously as well in NHPs. For instance, epilepsy, cerebral amyloidosis and cognitive changes due to aging can be found in these animals. Meanwhile, experimental inductions of Parkinson's disease, focal and generalized epilepsy, stroke and multiple sclerosis have also been performed.

4. *Gene therapy research*
   Gene therapy involves the delivery of specific genes to a patient as a treatment for disease. A carrier molecule called a vector must be used to deliver the therapeutic gene to the patient's target cells. Currently, the most common vectors used are viruses that have been genetically altered to carry normal human DNA. Research involving gene therapy on nonhuman primates often focuses on diseases caused by

single-gene defects. These include cystic fibrosis, haemophilia, muscular dystrophy and sickle cell anaemia.

5. *Virology research*
   Nonhuman primates are often the only species that can be used to study certain human viruses due to the relative homology and conservation of critical molecules used by viruses in their life cycle. Also, some of the natural viruses in these animals have their human counterpart. An example is the human immunodeficiency virus (HIV) and the simian immunodeficiency virus (SIV). The immune system of NHPs also closely resembles that of humans, in respect to development, genetics, function and anatomy. Recombinant HIV/SIV viruses (SHIVs) and hepatitis (B and C) have been widely studied in NHPs.

6. *Others*
   There are numerous other research uses for nonhuman primates.

   In 1996, a tuberculosis model using Philippine cynomolgus macaque was reported. Animals inoculated intratracheally inoculated with various doses of *Mycobacterium tuberculosis* developed either a rapidly progressive, fatal lobar pneumonia or a chronic, progressive localized form of pulmonary tuberculosis, similar to the disease in humans.

   Cynomolgus macaques have also been used extensively to study cocaine and alcohol abuse, including social factors associated with addiction. One study demonstrated that subordinate animals found cocaine to be more enforcing that did their dominant counterparts.

   Osteoporosis poses a major health problem for women, especially those at the post-menopausal stage. In macaques, the peak bone mass is at about 9 years of age. Procedures for measuring bone mass and density in women can be similarly applied to animals. Also, oestrogen deficiency (e.g., through surgical menopause) in cynos causes rapid bone loss that progresses for at least 18 months. This can be completely prevented with oestrogen treatment.

   The menstrual cycle and reproductive hormone profile of cynomolgus macaques are similar to women. Hence, they offer a good model for reproductive biology studies. Cynos are not seasonal breeders, thus allowing for reproductive functions to be studied year-round. This is unlike Rhesus macaques, which are seasonal breeders. Macaques have also been used to investigate breast and uterine cancer, particularly in relations to oestrogen exposure.

Future Advances in Primate Models

The use of NHPs in biomedical research is likely to increase as emphasis is placed on developing translational models for drug development that will be more predictive of efficacy in humans. Newer imaging modalities, such as PET-CT and functional MRI will be increasingly applied.

With the SARS epidemic in the recent past and the current threat of an avian influenza (bird flu) pandemic, more attention is being paid to emerging infectious disease research.

Nonhuman primates and other animal models are needed to study disease prevention, disease progression and effectiveness of possible treatment strategies.

While the research and medical community has long been interested in the possibility of transplanting organs from animals to humans (xenotransplantation), success has been limited. Many studies have used monkeys as the model recipient of pig organs in order to develop xenotransplantation strategies. With proper consideration of critical issues, such as rejection, zoonoses and ethics, the exploration of xenotransplantation will continue and NHPs will play a major role.

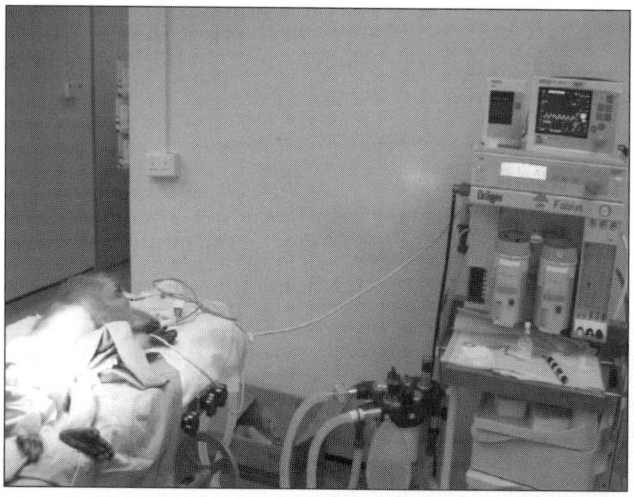

Fig 1.3.1: An anaesthetized cynomolgus macaque maintained on isoflurane, an inhalant anaesthesia.

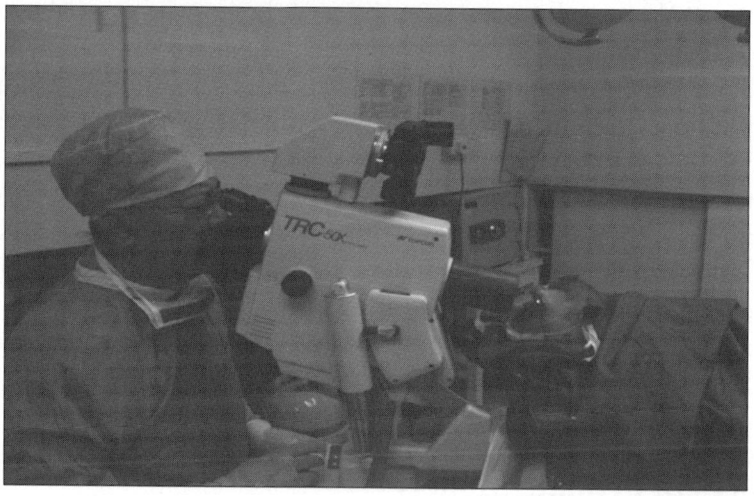

Fig 1.3.2: Ophthalmology procedure performed on NHP.

CHAPTER 2

REGULATORY CONSIDERATIONS IN THE USE OF ANIMAL MODELS

CHAPTER **2.1**

LAWS, REGULATIONS AND GUIDELINES FOR BIOMEDICAL RESEARCH IN SINGAPORE

Boon Theng Kuah

Over the years, Singapore has invested a great deal into developing the nation's capabilities in the area of the life sciences, focusing on biomedical research. We want to attract the best and brightest to Singapore as our partners in this relatively new field of endeavour, obviously in the hope that it will reap great rewards in the future. We also want to cultivate our own local expertise and experience to the point where we can maintain and sustain high standards of scientific work in an environment conducive of good and sound research. In order to achieve this, we must have a clear and strong framework for regulatory control and ethical oversight in the way in which research is conducted here, so that Singapore makes its mark as a place with excellent standards and quality work.

In Singapore, our framework of control over medical institutions that are involved in research is based on a licensing regime under the Private Hospitals and Medical Clinics Act. Under this statute, hospitals are issued with licenses to run medical/healthcare establishments and the license can include specialized activities that the issuing authority (the Ministry of Health) allows the hospital to conduct. In addition to the statutory provisions in the Private Hospitals and Medical Clinics Act (Chapter 248 of our Statutes) as well as other relevant legislation pertinent to healthcare services, from time to time guidelines and regulations may also be issued which the institutions must comply with. Failure to comply with these laws regulations and guidelines may of course jeopardize the institution's license to carry on its business and activities.

Law and Ethics

However the source of rules and guidelines telling us what is right and wrong in the area of healthcare in general, and biomedical research in particular, is never solely based on what the law prescribes. Healthcare and research standards must also be based on sound ethical principles. Long before any laws and regulations were implemented here, we have been looking to respected ethical codes such as the World Medical Association's Declaration of Helsinki, the International Conference on Harmonisation (ICH)/WHO Good Clinical Practice (GCP) standards etc, to guide us on what is right and wrong in human experimentation and research. In Singapore, we have our very own Singapore Good Clinical Practice Code that sets out some of the ethical principles of clinical research. And from time to time, our Bioethics Advisory Committee has issued Recommendations, which are relevant to healthcare and healthcare ethics, particularly relating to the field of human research.

It is of utmost importance for any responsible researcher in Singapore to know his or her ethical and legal duties and responsibilities, before embarking on biomedical research. The researcher must know what the prevailing rules and standards of conduct are, and not only that, he or she must further be conscious of new regulations and guidelines as and when they are updated and made applicable to research activity here.

In this chapter we will outline various sources of these legal and ethical obligations that researchers must remember, and also highlight some of the important principles to keep in mind.

Biomedical Research Bill 2003

If one were to make a search of our statutes and Acts of Parliament, you would not find very many statutory provisions dealing specifically with biomedical research. However in 2003, following disclosures of questionable practices relating to clinical research that had been conducted by the then Medical Director of the National Neuroscience Institute Professor Simon Shorvon and his team, a Committee was in fact set up to look into proposed legislation that would set out the legal duties and obligations in biomedical research, including criminalizing any violations. It was a clear signal that Singapore wanted to be tough on errant researchers who deviated from ethical norms and expectations, and wanted to clarify and spell out what the responsibilities were and what penalties would be meted out if researchers did not comply. The Committee eventually drew up the Regulation of Biomedical Research Bill of 2003 and this Bill was circulated to various interested parties so as to solicit their feedback and views. Once the consultative process was completed many waited with bated breath to see what final form any legislation would take. But the Act of Parliament was not to be. It appears that it was eventually withdrawn. One can only assume that there was a somehow a rethink on policy and principle, and the decision was made not to go the way of legislation and providing criminal sanctions to deal with research violations.

BAC Recommendations

On 16 September 2003 the Bioethics Advisory Committee issued a Consultation paper entitled "Advancing the Framework of Ethical Governance for Human Research" to invite views from professionals and the general public. Following this, on 23 November 2004 the BAC issued a Report entitled "Research involving Human Subjects: Guidelines for IRBs". This contained various recommendations on the ethical boundaries of clinical research. About a year later on 25 November 2005, the BAC followed up with a further Report entitled "Genetic Testing and Genetic Research", which followed an earlier Consultation paper issued seven months earlier on "Ethical, Legal and Social Issues in Genetic Testing and Genetic Research". This brings to four the number of Reports the BAC has issued from 2002 to 2005, which has provided recommendations on the ethical principles governing human research. This included other earlier papers on human tissue research, human stem cell research and reproductive and therapeutic cloning.

These Reports do not directly have the force and weight of the law, but they are certainly highly persuasive statements of what Singapore would regard to be reflective of our ethical principles and standards. Medical professionals who are involved in research and who act in a manner contrary to these established and persuasive sources of an "ethical code", would risk disciplinary action being taken against them for possible professional misconduct, or behaviour deemed to bring the medical profession into disrepute.

However the BAC Reports focus on human research, and it is clear their objective is in the impact of research on human subjects, material and data. Where then, do we look to to understand the fundamental principles that guide the conduct of animal researchers and institutions, which allow and facilitate such research?

Animal Research

It may be of interest to note that in the Regulation of Biomedical Research Bill of 2003, "biomedical research" was defined as follows:

"..any research—

> *(a) that is conducted for the primary purpose of increasing the fundamental knowledge and understanding of the physical, chemical and functional mechanisms of human life processes and diseases; and*
>
> *(b) that involves the use of human, animal or plant tissue.."*

By the above definition, biomedical research would include animal research, and the intention in drafting the proposed Bill was to place the standards and obligations of animal researchers on the same footing as those involved in human clinical research. This means that as a starting point, we do not set off by lowering standards when the subjects in research are animals and not humans.

Animals and Birds Act

Section 42 of the Animals and Birds Act (Chapter 7 of our Statutes) makes it a punishable offence to do any of the following:

- treat an animal cruelly.
- cause or procure or permit any animal to be used in such manner.
- neglect to supply an animal with sufficient food and water.
- wantonly and unreasonably doing or omitting to do any act, cause any unnecessary pain or suffering, or permit unnecessary pain or suffering to any animal.

The implications to animal researchers or any person handling such animals is clear. While the law may accept that some uses of animals in research may be necessary and legitimate in order to further knowledge and understanding in disease processes and treatments, since the use of animals in research would often cause pain suffering and distress to the animals, there must be reasonable justification, with a sound scientific basis, failing which it would just constitute an act of cruelty. Even if there is a legitimate and useful purpose in experimenting on animals, researchers who cause unnecessary pain or suffering to a research animal in the process would be guilty of an offence under Section 42. Responsible research should aim to minimize distress for the research animal. That is why considerations as to appropriate analgesia being used to relieve pain, the condition of the animals after commencement of the research (including the continued ability of the animal to eat and drink), the manner of transport, handling and housing the animal and even modes of euthanasia, are essential considerations for every research study. The IACUC is concerned about these issues because we do not want to see any research study run foul of Section 42 of the Animals and Birds Act. The IACUC also expects researchers to justify the number of animals used, since the excessive use of animals in experimentation would also be regarded as an unjustifiable act of cruelty under the Act.

Agri-Food and Veterinary Authority (AVA)

In Singapore our Agri-Food and Veterinary Authority (AVA) has oversight of anything involving the interest and welfare of animals and plants. The AVA is a statutory board that took over the responsibilities of the Primary Production Department. Their functions and powers are derived from the Agri-Food and Veterinary Authority Act, the key provisions which are highlighted below.

Section 11 of the AVA Act sets out the functions and duties of the AVA:

- *"to promote and regulate animal and fish health, animal welfare and plant welfare"*
- *"to develop, manage and regulate.. any other agri-food and veterinary centre or establishment"*
- *"to advise and make recommendations to the Government on matter, measures and regulations related to or connected with the agri-food and veterinary sectors.."*

Now the term "Agri-Food and Veterinary Sectors" over which the AVA has the duty and authority to oversee, is defined in Section 2 of the AVA Act to mean:

> *"..the sectors connected with.. animal health.. veterinary welfare...animal welfare.. protection of wild animals and birds, and includes the.. keeping, holding, catching, harvesting.. and trading of animals, birds.. reptiles, amphibians.. including their young, eggs and products derived therefrom.."*

One would be hard put to find something relating to animals that is not to be regarded as part of a "agri-food and veterinary sector" under this definition!

Powers of the AVA

The powers of the AVA over the agri-food and veterinary sectors are wide-ranging. Section 12 of the AVA Act provides that the AVA shall have power to do anything for the purpose of discharging its functions under the Act, or which is incidental or conducive to the discharge of those functions. Just to make doubly sure that we all understand the wide-ranging powers of the AVA in this, their area of jurisdiction, there are specific powers set out under the Act, and these include the following:

- To prescribe, regulate or implement measures and standards on any matter.
- To regulate, accredit or certify any person, company, corporation or organization to carry out any activity or service.
- To regulate, accredit, certify or implement any system, scheme or standard.
- To regulation, control or supervise any activity in any other agri-food and veterinary centre or establishment.
- To prescribe training requirements and provide training, conduct tests or award diplomas or certificates of proficiency.

In summary, under the AVA Act, the AVA both **prescribes** the applicable rules and guidelines affecting the care and use of animals, as well as **enforces** them. It is therefore both Regulator and Enforcer.

NACLAR Guidelines

One of the main initiatives of the AVA in relation to the use of animals in research, was to set up the National Advisory Committee for Laboratory Animal Research (NACLAR) in 2003 with a view to drawing up the National Action Plan for Animal Experimentation for Singapore. The declared purpose of the NACLAR Guidelines is to ensure the humane care and use of animals for scientific purposes, and to do so through

(i) the establishment of principles and guidelines governing care and use of animals for scientific purposes; and
(ii) the identification of responsibilities of investigators and research facilities using animals.

Under the NACLAR Guidelines, overall responsibility for the oversight and evaluation of all aspects of the institution's animal care and use programme lies with the IACUC. The

IACUC is also responsible for advising the Institution CEO of the steps required to maintain animal research facilities and to ensure that the program adheres to regulatory guidelines.

Responsibilities of Researchers under NACLAR

Amongst other things, the Guidelines also set out the responsibilities of researchers. Researchers are to ensure that:

- there is proper justification for the use of animals in the proposed research plan or protocol.
- they submit written proposals and obtain written approval before embarking on the research.
- they refine the study design and techniques in order to minimize distress to the animals.
- they give adequate consideration to the living conditions of the animals.
- they use the best scientific techniques and ensure competence in the procedures being performed.

Researchers should also take note that the NACLAR Guidelines also provide rules as to procurement of the animals used in research. They should familiarize themselves with these rules before drawing up their plans for obtaining the animals they need.

The Guidelines are as instructive to the animal research facilities as they are to researchers. Amongst other things, it sets the standards and requirements for the animal research facilities keeping, using and facilitating the use of such animals, including rules as to the management of animals in breeding and holding areas.

Important Principles

In my view, the NACLAR Guidelines as well as the prevailing laws and ethical standards in Singapore, provide us with several key principles that all those involved in or intending to participate in animal research must bear in mind.

1. Those who use animals for scientific purposes have a legal, moral and professional duty to treat the animals humanely and consider the animals' welfare when planning and conducting experiments.
2. As part of that legal and ethical duty, investigators must ensure that the medical and surgical techniques used are consistent with the principles of good practice and scientific knowledge in laboratory animal veterinary medicine.
3. However investigators are often not experts in veterinary care. They should therefore consult with veterinarians whenever prudent, particularly when adverse events occur, so that appropriate veterinary care and treatment are carried out and made available.
4. Investigators have direct responsibility for all matters related to the welfare of the animals under their control and this responsibility extends to all facets of the care and use of animals in projects approved by the IACUC.

5. Primary investigators are responsible for the standard of animal care and use by all other persons involved in the research. They should therefore ensure that the extent of supervision is compatible with the level of competence of each person in the team and the duties assigned to each member of the team are appropriate to that person.

With time, the emphasis on high ethical standards in biomedical research can only grow, and our rules and guidelines become more developed. Unethical practices bring the institution and the nation's efforts in developing and promoting biomedical research into disrepute, and Singapore cannot afford to allow this to thwart our growth and progress in this area.

The starting point is with the recognition that research subjects, whether they are humans or animals, are to be protected and respected. Animals are valuable, and we should make their welfare our priority. Our laws, regulations and guidelines relating to animal welfare and use for scientific purposes are in fact very well developed, with the implementation of the NACLAR Guidelines providing much clarity and purpose in this area. Institutions and researchers must ensure a high level of accountability for their research activities, including to their respective IRBs and IACUCs, as well as to the AVA and ultimately, to the law. I am sure we will continue to see further refinements in the way our laws and practice standards will evolve in the future.

CHAPTER **2.2**

THE FUNCTIONS OF THE INSTITUTIONAL ANIMAL CARE AND USE COMMITTEE

Pierce Chow

Established scientific and educational institutions commonly have oversight bodies to govern the use of animals for scientific research within the institutions. The composition, functions and responsibilities of such bodies will vary depending on existing (or absence of) local legislations. Such oversight bodies serve to protect the institution and assure the public that biosafety guidelines are followed, acceptable standards of animal research are maintained and that the treatment of animals within the institution is appropriate and humane.

These bodies are known by various names such as Institutional Animal Ethics Committee and Institutional Animal Review Committee. In the North American continent and in Singapore, these oversight bodies are known as the Institutional Animal Care and Use Committee (IACUC).

The regulations pertaining to the composition, practices and responsibilities of the IACUC in Singapore are well defined. They are also very similar to those in many other countries where guidelines and legislations on research using animal models are well established. The discussions and descriptions of a functioning IACUC in this chapter are based on processes and practices at the Singapore Health Services (a biomedical institution), which has the IACUC with the longest established history in Singapore.

The IACUC in Singapore

Any institution in Singapore intending to use animals for scientific purposes will require an appropriate license from the Agri-Food and Veterinary Authority (AVA) of Singapore. Only the following types of institutions may qualify for such licenses: institutions of higher learning, statutory boards and biomedical organizations and businesses (both public and private). Issuance of this license is subject to these institutions fulfilling requirements specified by the Animals and Birds Act (Care and Use of Animals for Scientific Purposes) (http://www.ava.gov.sg/AnimalsPetSector/CareAndUseAnimalsForScientificPurp/index.htm). These requirements include compliance with the Guidelines of the National Advisory Committee for Laboratory Animal Research (NACLAR) and inspection of its facilities by AVA. The existence of an appropriately appointed and fully functional IACUC is also a pre-requisite for such a license.

NACLAR Guidelines were adapted from the best practices of Australia, Canada, New Zealand, the US, and various organizations including the Council for International Organisations of Medical Sciences (CIOMS) and the European Convention for the Protection of Vertebrate Animals Used for Experimental and Other Scientific Purposes (1986). NACLAR Guidelines are in many ways similar in spirit and content to the guidelines issued by the National Research Council in the US. The Guidelines defines scientific purposes broadly to include the use of animal in teaching, field trials, environmental studies, research, diagnosis, product testing, and the production of biological products. Details pertaining to the Institutional Animal Care and Use Committee (IACUC) are also described.

Licenses to use animals for scientific purposes are issued by the AVA to the CEO (or equivalent individual) of the organization, and the licensee is expected to enforce compliance with the Guidelines in his institution. Infringement of the rules are subjected to penalties under Section 19 of the Animals and Birds (Care and Use of Animals for Scientific Purposes) Rules which includes a fine up to S$10,000 or imprisonment for up to 12 months or both. In addition, the animal research facility license may be revoked or suspended under Section 62 of the Animals and Birds Act.

In order that the institution continues to maintain such a license, the CEO, in practice depends on the IACUC to ensure that the institution complies with NACLAR guidelines. Enforced self-regulation is expected on the part of the institution. The IACUC is responsible to the CEO for oversight of the animal care and use programme in the institution and advises the CEO on how compliance with the Guidelines can best be achieved. (The term "programme" in this instance refers to all relevant policies and protocols in the institution pertaining to the care and use of animals in research and training as well as the institutional philosophy underpinning the protocols). In addition, the IACUC also has specific responsibilities (described below).

Authority, Composition and Qualifications of the IACUC

Consistent with its responsibilities to the CEO, the IACUC is appointed directly by the CEO and reports only to the CEO (Fig. 2.2.1). This allows the IACUC the necessary independence to enforce regulations without undue administrative hindrances. The composition of the IACUC is strictly regulated and comprises of at least five members in four separate categories. While the IACUC can certainly have more than 5 members (and this is necessary in large institutions), no member may be in more than one category. Any category may however have more than one person.

The 4 categories are:

1. *Veterinarian*
   Defined as a person with suitable qualifications in veterinary science and licensed by AVA to practice veterinary medicine in Singapore. He is also expected to have received appropriate training in or experience in laboratory animal science and medicine. He must, by definition, be the official Institutional Veterinarian and has direct or delegated responsibility for activities involving animals in the research facilities. While the committee can have more than one veterinarian, only one can be the official veterinarian in the committee.
   The appropriate training of the Institutional Veterinarian is defined under Article 2.4 of the NACLAR Training Guidelines.

2. *Scientific Person*
   This is defined as a person with substantial recent and appropriate experience in the use of animals for scientific purposes. In practice this usually entails the possession of a relevant post-graduate qualifications for example M.Sc., PhD, and active participation in ongoing research using animals.

3. *Non-Affiliated Person*
   A person not affiliated in any way with the facility. By definition this person may not be a member of the immediate family of a person working in the institution.

4. *Non-Scientific Person*
   A person who represents the interests of the general community in the proper use and care of animals. By definition he/she should not be a user of animal for scientific purposes. Suitable persons would include clergy, lawyer, ethicist etc.

In order to avoid undue partisan influence on IACUC decisions, not more than three voting members of the IACUC may be from the same department or unit in the institution. For the committee to arrive at fair judgements and decisions, members are expected to come from diverse backgrounds. Nominal compensation is permissible and the IACUC may consult external consultants although only IACUC members may make the relevant decisions.

The Chairman of the IACUC is appointed directly by the CEO and the regulations stipulate that he may not be the Institutional Veterinarian. In practice the chairman should be a senior person within the organisation with the necessary authority to enforce decisions. In cases where the chairman is an external person, the CEO has to ensure that he be vested with sufficient authority.

The Training of IACUC Members

The Guidelines stipulate that at least 50 % percent of the members of the IACUC must have undergone formal training in IACUC work (for example, participation in formal IACUC training workshops). On-line training supplements this training and keeps the members updated on all IACUC trends and developments.

Suggested modules for training include the following:

- ARENA IACUC 101 Workshop.
- Orientation Module – Programme & Education Training for New IACUC Members.
- Recommended Continuing Education Module.
- Internet On-Line Training.

Specific Responsibilities of the IACUC

The main responsibility of the IACUC is to ensure that the institution does not lose its license to use animals for research and training by complying with the Guidelines. It does this by annual inspection of the animal facilities (hardware), review of the animal programs (software), review and approval (or rejection) of proposals to use animals, and accreditation of individuals to use animals in the institution. The IACUC reports to the CEO all relevant deficiencies and make recommendations to rectify them. The reports of the IACUC are accessible by AVA.

Specific responsibilities of the IACUC laid down by the Guidelines are:

1. *Annual Evaluation of the Institution's Animal Facility*
   In this respect the IACUC is in an advisory role to the CEO and in its annual report, the IACUC advises the CEO on the institution's compliance. It suggests plans and schedules for correcting deficiencies necessary to either maintain or achieve compliance, and makes recommendation regarding the institution's animal program, facilities or training of personnel. The CEO has the responsibility of making the appropriate decisions in the light of the recommendations. This is "enforced self regulation" in practice.

2. *Six-monthly Review of Programmes*
   All programmes for the care and use of animals carried out in the facility must be reviewed.

3. *Reviews Proposals for Research Projects*
   This is carried out independent of the CEO who may not overrule IACUC decisions. The institution is however not required or obligated to conduct a research activity approved by IACUC. The institution may also subject research protocols to additional institutional review (e.g. by department head, bio-safety committee, etc.) and may seek assistance when necessary from other authorities e.g. National Environmental Agency (see details below).

4. *Maintains Register of Approved Projects*
   This must be made available for inspection by AVA.

5. *Certifies Individuals Qualified to Use Animals for Research*
   In larger institutions, the IACUC is actively involved in the running the courses prescribed by NACLAR Guidelines (the Responsible Use and Care of Animals course) to certify individuals who may be involved in using animal for scientific

purposes. In smaller institutions, the IACUC has the responsibility for ensuring that all individuals using animals are suitably certified.

6. *Reviews and Investigates Animal Welfare and Biosafety Complaints (see details below)*

7. *Monitors Compliance*
   Many methods are available to IACUC and these include the tracking of animals (paper or electronic), the use of compliance specialists, the annual inspection, retrospective reports of adverse events and review of publications.

   In practice the most important resources to the IACUC in this respect are the veterinary and research staff (eyes and ears). Motivated staff and researchers are is the most important and practical ways of continual monitoring.

Review of Proposals for Research Projects

The researcher comes into contact with the IACUC mainly through the process of applying for approval of research projects. The detailed process of research proposal evaluation including criteria for approval will be covered separately in Chapter 2.3. The process of application is discussed here (Fig. 2.2.2).

The Principal Investigator submits a soft copy of the stipulated application form (which is obtained online from the intranet) by email and a signed hard copy of the same form to the IACUC secretariat. The secretariat will distribute soft copies of the application form to all IACUC members at least one week before a scheduled IACUC meeting for preliminary review by members.

Hard copies of the completed application form will be distributed to members at the IACUC Meeting, which is held every three to four weeks. Prior to the meeting the chairman appoints a primary reviewer for each application. The primary review makes a detailed study of the application and will make a presentation of the application to the other members during the IACUC meeting. Other members of the IACUC will be asked for their views, comments and endorsements.

A research project application is approved if a quorum (50 % of committee) is present at the Meeting and if more than 50 % of the quorum votes in favour. Approval must also require the presence of member(s) in either category (3) – member representing general community and/or (4) – member whose primary concern is in a non-scientific area. Adhoc consultants may be consulted to assist with expert opinions but the final decisions may only be made by IACUC members.

After reviewing an application for a research project the IACUC will make one of 3 decisions namely: approval, approval with modifications or rejection of the application. The IACUC frequently allows researchers to reply to queries from reviewers and holds the decision until outstanding issues have been clarified. Research proposals found to fall short of the Guidelines will not be approved and the IACUC may suggest amendments and modifications to allow it to comply with the Guidelines in order to secure approval. The identities of reviewers are frequently made known to investigators and it is not uncommon for reviewers to directly contact applicants before the IACUC meets in order to clarify issues and make suggestions.

All decisions pertaining to the application will be conveyed to applicants in writing. In cases where the application is not approved, the investigator will be informed by email of the specific reasons for non-approval. Replies from investigators will be assessed at the next meeting and if found acceptable, approval of the application will be conferred. All approved proposals and accompanying documents will be filed in the IACUC secretariat for a period of at least 3 years beyond the completion of the project. Approved protocols must be reviewed every year and the investigator is expected to submit a brief report.

Investigators may only begin with their experimentation after written approval from IACUC.

An IACUC member may not participate in the approval process if there is conflict of interest such as those, which can arise under the following circumstances:

- The member is himself an applicant.
- The member has competing research of his/her own.

Research is a dynamic and not a static process and frequent amendments to the protocol may be required in the course of the research. Such applications for amendments to the protocol must receive IACUC approval before they are effected or they will be considered breaches of protocol. Common amendments are:

- Changes in study design or increase in number of animals.
- Changes in drugs used.
- Change in **personnel** carrying out the research.

If breaches to a previously approved protocol occur, this may result in the project being suspended or the approval being formally withdrawn. On its part the IACUC makes all attempts to review amendments quickly so as not to hold back research and decisions are frequently made online to achieve this.

Investigators are required to inform the IACUC in writing when projects are completed or discontinued and the general outcome of each project must be made known to the IACUC.

Investigations of Animal Welfare and Biosafety Complaints

The IACUC has a duty to establish channels for feedback on grievances and other complaints about breaches in appropriate animal care and use within the institution. Individuals who identify deficiencies in animal care and use in the institution or biosafety concerns may report their concerns to any member of the IACUC. Appropriate channels for such complaints (such as contact numbers) must be easily visible within the animal facilities.

When a complaint is received, the IACUC will verify the stated concerns and where appropriate establish guidelines for effecting corrective measures. All complaints will be formally acknowledged. The IACUC will then meet to decide if the complaint has sufficient substance to merit further investigation and to decide on the investigational steps. The relevant individual will be informed and given the opportunity to explain the circumstances.

The results of investigations carried out by IACUC will be made available to all parties involved. If violations of rules or regulations are identified, the IACUC may suspend activities involving the animals and revoke the right of the researcher to use animals and the CEO will be informed. External authorities will be informed as appropriate.

No employee of the institution should face discrimination or be subjected to reprisal for reporting animal welfare concerns.

Central Role of the IACUC

The IACUC plays a central role in maintaining high standards of research in the institution. Members of the IACUC should be chosen carefully as the IACUC is a significant determinant of the research culture within the institution. It is also the main interface between the users of the facility (investigators) and institutional administration (the licensee) and play important roles in ensuring biosafety and in providing assurance to the public of appropriate care of animals used for research in the institution.

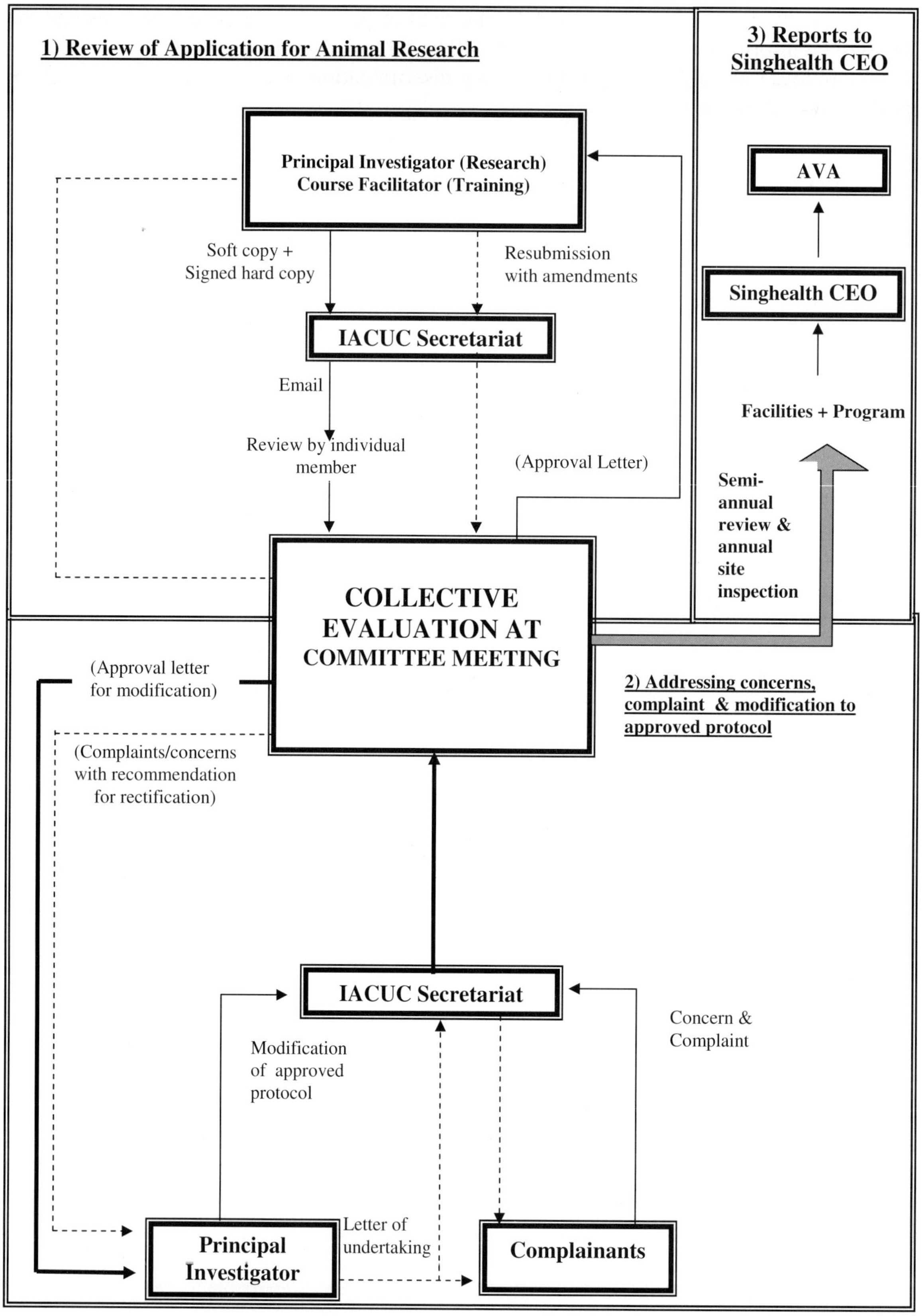

Fig 2.2.1: Flowchart of Singhealth-IACUC.

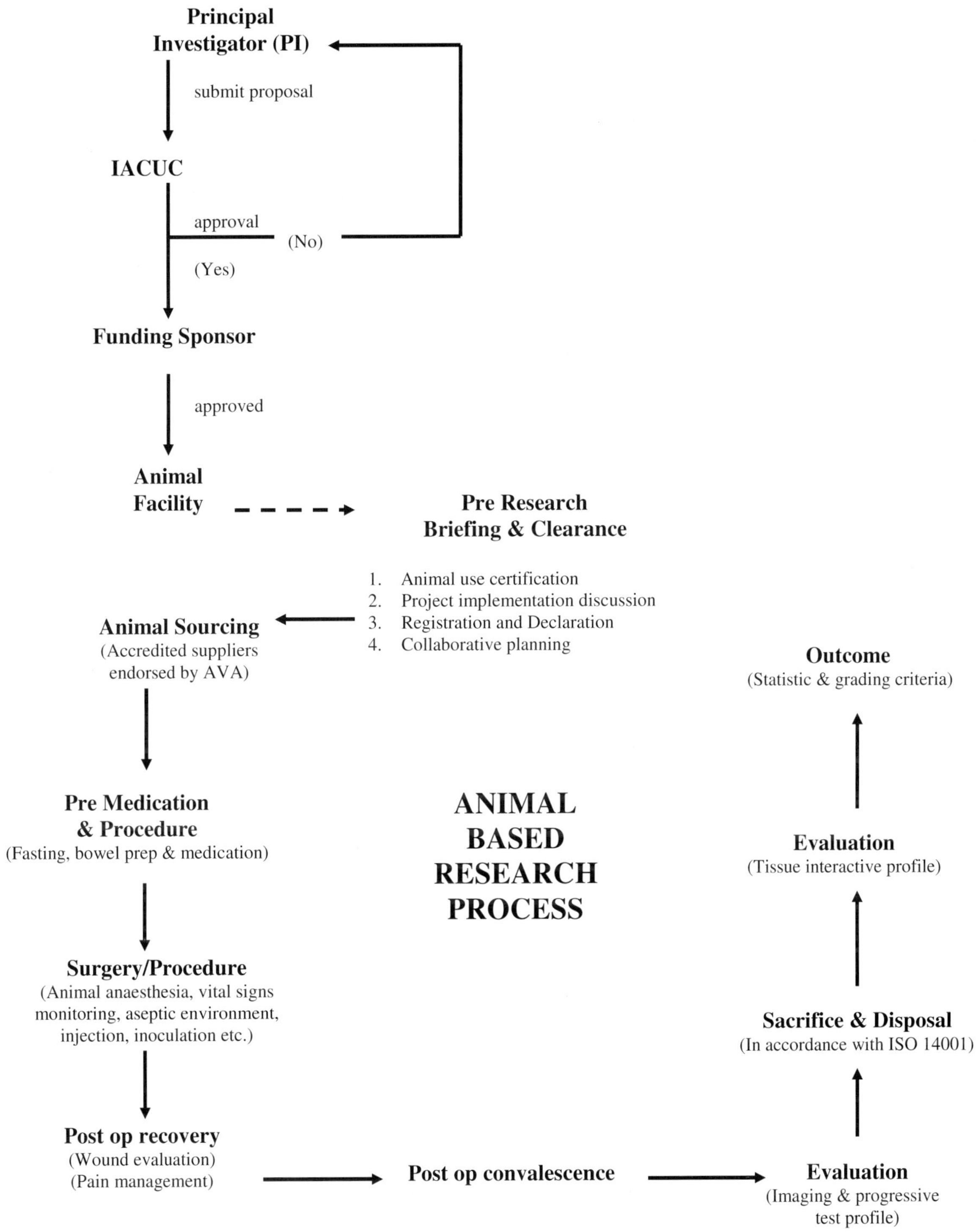

Fig 2.2.2: Workflow for animal research.

CHAPTER **2.3**

RESPONSIBILITIES OF PRINCIPAL INVESTIGATORS AND RESEARCH PROTOCOL EVALUATION

Hock Soo Ong

Responsibilities of Investigators

1. *Direct and Primary Responsibilities*

   The investigators of a research project are *directly* and *primarily* responsible for all matters relating to the animals that have been allocated to them. The investigators are *ethically*, *professionally* and *legally* responsible for these animals. They are to ensure that these animals are maintained and manipulated in accordance to good practice guidelines and sound scientific principles.

   Animals used must be regarded as sentient. Hence, they must be handled and manipulated humanely according to ethical principles as laid out in the "Guidelines on the Care and Use of Animals for Scientific Purposes" (NACLAR). The investigators have a professional obligation to their institution and the scientific community to ensure that their privilege to do animal research is not jeopardised by callous and careless attitudes and practices. Under the Animal and Birds Act, the Agri-Food and Veterinary Authority (AVA) has vested powers to withdraw the license of an institution to conduct animal research and/or prosecute individuals for failure to comply with the guidelines specified.

   The role of IACUC is to ensure that all projects involving the use of animals comply with the guidelines and requirements as prescribed by the AVA.

2. *IACUC Approval*
   It is the onus of the investigators to submit an application to the IACUC prior to starting any project involving animals. Investigators must not begin work before receiving written approval from the IACUC.
   After approval has been granted, investigators must ensure that they adhere to the protocol and any other requirements requested by the IACUC. The period of accountability starts from the time the animals are allocated and ends only with proper euthanasia and disposal of the animals after use.

3. *Planning Projects and Conduct of Experiments*
   The investigators have the ethical responsibility to ensure that the principles of the 3 Rs (Refer to Chapter 2.4) are applied in the planning of the projects and observed in the conduct of the experiments. In addition, the investigators must not deviate from the protocol submitted. IACUC approval must be obtained prior to making any significant changes to the approved protocol. Examples of significant changes where IACUC approval are required are:

   - change in objectives.
   - change from non-survival to survival surgery.
   - change in invasiveness/discomfort to animals.
   - change in personnel.
   - change in species, number of animals.
   - change in duration, frequency and number of procedures.
   - change in methods of euthanasia.

4. *Training*
   The investigators are responsible for reporting to the IACUC (via the animal research proposal application) that they and their staff have sufficient qualifications and training in the procedures to be performed on animals. Completion of the "Responsible Care and Use of Laboratory Animal" course is the minimal requirement for any investigators involved in animal research. "Advanced and Special" courses may be needed for more complex research programmes such as special training for handling nonhuman primates.

IACUC Criteria for Evaluation of Research Project Proposal

IACUC will evaluate submitted research protocol based on a checklist of requirements. Incomplete details and information will delay approval of proposal.

1. *Study Objectives*
   The scientific objectives must justify the distress and discomfort caused to the animals. The study objectives must be relevant. It must be of value in contributing to the advancement of knowledge for the good of society and animal or human health.

2. *Research Duplication*
   Investigators are required to indicate that the proposal is not a repeat of previously reported experiments. If similar experiments had been performed, investigators have to explain the refinements in the current proposal and justify the duplication.

3. *Selection of Appropriate Animals*
   Animal chosen must be of a species appropriate for the scientific purpose. This is usually based upon anatomical, physiological and other characteristics that suit the scientific objectives and fulfil the need for clinically valid results. Bred animals are preferred to wildlife. Submission of references and source of animals to be purchased will facilitate evaluation. Animals must be obtained from AVA accredited sources. (Refer to Appendix 1).

4. *Number of Animals Requested*
   Number of animals required must be justified based on scientific and/or statistical considerations. The number of animals requested should be the minimum required for robustness of scientific data and statistical significance. Prior consultation with biostatistician will be helpful.

5. *Transportation of Animals*
   Transportation causes stress to animals because of confinement, sudden movements, noise and changes in the environment and personnel. Such stress may become a significant experimental variable. If not handled properly the animal can also cause injury to itself or its handlers. Hence, transportation of animals between approved institutions should be kept to a minimum. Transfer of animals from approved institution to an institution that is not approved by IACUC is not allowed.

   Containers used must be escape and tamper proof and should be protected from sudden movements. The transfer of genetically modified animals between approved institutions should be in accordance with the guidelines of the Genetic Modification Advisory Committee (GMAC).

6. *Animal Procedures*

   - **Injection and Inoculation**
     Procedure that involves injection or inoculations of pharmacological agents should come with the following information:

     - Name of the injectable substance.
     - Dose and volume of substance to be injected. Dose should be expressed as weight of substance over animal weight, for example, mg/kg and volume of injection in ml or μl.
     - Site and route (intravenous, intramuscular, intradermal, subcutaneous or intraperitoneal) of injection.
     - Schedule of injection.

- **Blood Withdrawal**
  Site, frequency and volume of blood to be aspirated must be specified (Refer to Chapter 3.2). As a guide, blood withdrawal is limited to 1 % of lean body weight every 3 weeks, for example 2 ml for a 200 g rat or 0.06 % of the lean body weight daily for up to three weeks. Animal welfare is the prime consideration in blood sampling. However, investigators should realise that excessive blood sampling will affect the physiological response of the animal and hence the quality of data collected from such animals.

- **Methods of Restraint**
  Physical restraint devices such as rabbit or rodent restrainers, swine slings or monkey chairs are useful for certain non-painful procedures. When physical restrainers are used they should provide animals with the opportunity to assume their normal postural adjustments. Physical restraint devices should be specifically designed for the particular species and used only when other means are not possible or impractical. Investigators should observe the following guiding principles in the use of physical restraint devices:

  - Restraint devices are not to be used as normal housing for the animals.
  - Animals to be placed in restraint devices should be given time and training to adapt to the equipment and personnel.
  - The period of restraint should be the minimum required to accomplish the research objectives.
  - When restrained, the animal should have access to food and water at regular interval. If an animal is restraint for prolonged period, consideration must also be given to its need of exercise to prevent muscle atrophy.
  - Animals that are restrained must be monitored so that there is no inadvertent escape from the restraints or injury cause to itself in the process of manipulating itself out of the restraint.
  - Veterinary care should be provided if lesions or illnesses associated with restraint are noted.

The use of chemical restraints (pharmacological methods for restraining) should always be considered in painful or prolonged procedures.

- **Animal Identification**
  Investigator should indicate the appropriate type of identification for their study animals. Identification methods are based on the size and available marking site of the animals. Available methods include tattoo for pigs, goats, monkeys and rodents. Electronic microchip for rabbits and monkeys, physical marks such as ear punch for rat and mice or by cage card labelling. The method of identification should be reliable and cause the least stress or possible injury.

  Proper animal identification is important for accountability and accurate data collection.

- **Level of Pain and Distress**

  Whether procedures cause pain or distress in an animal is sometimes difficult to assess. The IACUC considers procedures painful if they produce a stimulus that humans would experience as painful, that elicit escape behaviour in animals, or that is capable of damaging tissue. The IACUC holds that procedures cause distress if they involve any disruption of physiologic equilibrium manifested by abnormal or maladaptive behaviour.

  In the research project proposal investigators are required to classify their project's procedures into three categories:

  - Minimal, transient or no pain or distress. Examples include injections, ear tagging, restraint of less than 12 hours, and food deprivation for up to 48 hours using well-nourished, healthy adult animals.
  - Pain or distress relieved by appropriate measures. Examples include surgery, retroorbital blood collection and trauma.
  - Unrelieved pain or distress (not relieved by anaesthetics, analgesics, or tranquilizers). Examples include experimental induction of a debilitating disease.

  A project may involve procedures in more than one category in which case the project is classified in the most advanced category.

  For each procedure and its anticipated resultant effects, the proper and appropriate relieving measures (analgesia, anaesthetics, tranquilizers) must be clearly described.

- **Duration of Study**

  The duration of a project must not be longer than that necessary to achieve the scientific objectives. Projects that cause significant pain and distress to animals must be as brief as possible. Any decision to observe such animals over a long period must be based on the well being of the animal.

- **Major Surgery**

  Major surgery is defined as procedure where a body cavity is exposed or penetrated or where it produces substantial impairment of physical or physiological functions.

  In the research setting, two types of major surgeries are performed. *Non-survival surgery*: surgical procedures are performed on the animal under anaesthesia and the animal is subsequently euthanized at the end of the procedure without reversal from anaesthesia. *Survival surgery*: surgical procedures are performed on the animal under anaesthesia and the animal is subsequently reversed alive and maintained for data collection. Surgical procedures under this category need to be performed under stricter conditions.

  Description of the surgical procedure should include:

  - Preoperative preparations: premedication, bowel preparation, fasting and animal health checks.

- Perioperative preparations: prophylactic medications (antibiotics, heparin, etc), skin preparation, vital signs monitoring, parenteral access, wound closure technique.
- Postoperative preparations: post-op pain control and medications, infection control measures, clinical assessment and other treatment protocol.
- Personnel performing the surgery. This is usually a surgeon, scientist or veterinarian with surgical experience in laboratory animals.
- Facility where surgery will be performed. The facility may be species-specific or procedure specific. Surgery can only be performed in an IACUC approved facility.

- **Multiple Major Survival Surgery**

  Multiple major survival surgical procedures on a single animal is discouraged unless it is for the following reasons:

  - Procedures are related component of a research project e.g. the creation of a surgically induced animal model to be followed up by interventional surgical research.
  - Conservation of scarce animal resource.

  If multiple major surgery is approved, the IACUC will pay particular attention to animal well-being through continuing evaluation of outcomes. Multiple surgery as a means of cost saving alone is not an acceptable reason.

7. *Humane Study Endpoints/Experimental Endpoints*

   Death as an experimental end-point is generally unacceptable and must be justified. For all but the most minor manipulation, the investigator should develop humane experimental end-points. These are criteria that are used to judge when an animal should be euthanized or removed from the study for treatment, in the interest of animal welfare. End-points earlier than the moribund condition should be used. As a general guide, animals should be euthanized when:

   - They have lost more than 20 % of their body weight or more than 10 % in 24 hours.
   - A tumour grows to more than 10 % of the animal weight or when the tumour ulcerates or abscesses.
   - Body temperature falls below a preset level, which is predictive of death.
   - Animals self mutilate.
   - Animals become obviously incapacitated and not able to feed, rest or perform normal activities.

8. *Method of Euthanasia*

   The method of euthanasia is based on the species, size of the animal, the scientific objectives of the experiment and its ability to quickly and painlessly produce a loss of consciousness and death. In general the acceptable method of euthanasia should have the following characteristics:

- Causes minimal pain, distress, anxiety or apprehension.
- Minimal delay until unconsciousness.
- Reliable and irreversible.
- Safe for handling personnel.
- Compatible with scientific requirement and purpose.
- Compatible with species, age and health status of animal.
- Drug availability and its human abuse potential.

Chapter 3.8 lists the different methods of euthanasia. Acceptable methods are recommended methods because of its reliability and irreversibility. Conditionally acceptable methods can be used only if there are applied together with another modality such as cervical dislocation of a rodent that is still under anaesthesia. Unacceptable methods are not approved because it does not meet the characteristics listed above.

9. *Carcass Disposal*
Dead animal carcass must be disposed in a manner that does not compromise occupational health and safety guidelines. It is recommended that all carcasses be deposited into a ziplock bag, sprayed with disinfectant and then placed into a yellow coded biohazard bag which is cord tied and discarded into a designated covered container. Disposal contractor will collect waste bags on the same day for incineration at Tuas Incineration Plant.

10. *Hazardous Materials*
IACUC will pay particular attention to proposals employing potentially hazardous materials including:

- **Radioactive Substances**
All individuals using radioactive materials must be registered with Radiation Protection Inspectorate (RPI), Health Sciences Authority. Investigators must specify dosage and schedule of radioactive substance used. Any exposure value outside the animal cage that exceeds 2 millirem/hour must be shielded by 1" thick perspex for beta rays and lead shield for gamma rays. Only short half-life radionuclides will be considered and MSDS have to be submitted. (Appendix 2).
The use of these substances is subject to regulations of Institutional Radiation Safety Committee.

- **Biological Agents/Recombinant DNA**
The uses of these agents are subject to regulations and approval by the Institutional Biosafety Committee. The IACUC will not consider such application until a written approval is obtained.

- **Hazardous Chemicals or Drugs**
  Activities involving hazardous chemicals require procedures for

  - Chemical storage and disbursement.
  - Dosage preparation and challenge procedures.
  - Waste management and disposal practices.

  Such chemicals may find their way into feed, faeces or urine hence adequate and appropriate safety practices, containment and facility safeguards must be available. Proposals submitted to IACUC must include sufficient documentation including MSDS to assess the adequacy of precaution to control exposure of personnel to hazardous agents involved in animal experiments.

CHAPTER **2.4**

THE 3R'S, RESEARCH VARIABLES AND THE USE OF ALTERNATIVES

Hock Soo Ong

The use of animals in research, training and teaching has important ethical and legal implications. In some countries, it is even a political issue. The issue relates to the moral value that one places on humans relative to animals. Views on the use of animal ranges from one of "animal rights" to that of animals as a resource for human use. However, most agree that animals in different classes or order differ in its significance and value. For example, a nonhuman primate differs from the Drosophila — one has neurological systems similar to human and display advanced social behaviour, the other does not. This has led to movement and legislation requiring humane treatment of certain categories of animals.

The Guiding Principles published by National Advisory Committee on Laboratory Animal Research (NACLAR) cover all live fish, amphibians, reptiles, birds and nonhuman mammals. It extends to all aspects of their care and use — teaching, field trials, environmental studies, research, diagnosis, product testing and production of biological products.

Alternatives that refine existing methods to minimize distress to animals, reduce the number of animals required for an experiment, or replace the use of whole-animal must be considered. These principles of humane experimental techniques are encapsulated by Russell and Burch in their book *The Principles of Humane Experimental Technique*, published in 1959. They classified humane techniques under the headings of replacement, reduction, and refinement — now commonly known as the three Rs.

Replacement

Replacement could be achieved by substituting one of the following for commonly used animals in research:

1. *Non-living systems*
   These are physical, mechanical, chemical, mathematical or computer simulation techniques that may complement or replace the use of living animals. Physical and mechanical models are widely available for training in cardiopulmonary resuscitation, minimally invasive surgical techniques and other clinical skills such as CVP insertion, venepuncture and trauma management. Chemical and radioimmunoassay techniques may provide alternative to animal testing. Mathematical models and computer simulations are possible as knowledge of biochemical processes becomes defined in quantitatively and mathematical terms. Computers cannot generate new biological information but they simplify the analysis of vast amounts of data and test hypothesis. For example, computer can be used to scan thousands of chemicals for a certain cytotoxic activity prior to animal testing.

2. *Living tissue systems*
   These are *in vitro* methods that utilize organ, tissues or cell culture techniques. Techniques are now well developed to culture cells, tissues, and organs of different species including that of humans. Advantages of this technique are that cells and tissues can be maintained in a defined, controlled environment, they may retain the differentiated functions that exist in the whole body system and they provide a rapid and less expensive means of evaluating physical and chemical agents. In fact, the use of cell and tissue cultures has allowed the discovery of information that would not have been obtainable from research on more complex systems. Limitations of this technique are that cultured cells may lose their differentiated function, cultures may not reflect the *in vivo* response that results from complex tissue and organ interactions, cells in culture may mutate and a particular observation may be confounded by infection of the culture with undetected pathogen.

3. *Microorganisms and invertebrates*
   Microorganisms (such as yeasts, bacteria) and invertebrates have less complex system that may provide useful knowledge of the underlying processes that may be clues for disease and disorder in higher animals. The bacterium (*Escherichia coli*), yeast (*Saccharomyces cerevisiae*), roundworm (*Caenorhabditis elegans*), and fruit fly (*Drosophila melanogaster*) are well-studied models, which are useful platform for genetic research. These organisms can be generated quickly and in large quantities, which are useful properties for such research.

4. *Non-mammalian vertebrates or animal lower on the phylogenetic tree*
   Studies using the giant axon of the squid and mesentery of the frog have been fundamental to our understanding of neuroscience and microvascular physiology. Tests for carcinogens are classically performed on rats and mice. However, small fish species have proven to be useful environmental sentinels as well as versatile test animals in toxicity and carcinogenicity bioassays. Advantages are their low

maintenance cost, low background incidence of tumour and ability to breed in large numbers rapidly.

Although lower organisms are excellent models for the study of certain basic life processes, interspecies transfer of information must be approached with caution and requires validation in higher animals.

Reduction

The number of animals required should be the minimum to achieve scientifically valid results. However, this principle should not be at the expense of greater distress and pain to the smaller number of animals. In line with this principle unnecessary duplication and repeats of experiments involving animals are not allowed. Number of animals used can be reduced without compromising the robustness of the data desired. This can be achieved by observing some of the following measures:

1. *Rational selection of group size*
   A pilot study is a useful basis for sample size estimation. A small group of animals can be studied to test feasibility of a hypothesis and estimate the variability of data before embarking on a full-scale study involving a larger group of animals.
   Sample size is often dependent on the magnitude of the effect. Application of statistical methods or consultation with a biostatistician is essential and helpful.

2. *Careful experimental design*
   Careful experimental design ensures reliable data and avoids repeat experiments because of poor planning. Some examples of design factors that can reduce required animals are:

   - Ensure research protocol covers all intended measurable parameters to avoid repeat experiments.
   - Standardize procedures to minimize variables.
   - Control animals given innocuous substance, such as saline, can be shared with other investigator.
   - Performing several terminal procedures per animal.
   - Animals euthanized by one investigator can be used for tissue needed by another.
   - Good postoperative care and infection control measures minimize loss of animals
   - Take measures to avoid unintended breeding.
   - Good record keeping and maintenance of data avoid loss of data and repeat experiments.

3. *Minimizing non-experimental variables*
   Many non-experimental variables may affect the results of an animal experiment. It is important to be aware of the factors and minimize their presence for robust and accurate data.

Choice of Animal

Animals chosen must be of appropriate species and quality to answer the scientific question. In general, bred animals are preferred to wildlife. Proper choice of animals minimizes genotypic and phenotypic variables in the physiological response of the animal.

- ***Specific Pathogen Free (SPF)***
  Pathogens can affect research by causing disease and death and thus confound interpretation of results or invalidate the study. Specific Pathogen Free (SPF) rats, mice and rabbits are available and these should be used whenever possible.
- ***Anatomical Compatibility***
  The primary aim of animal research is clinical application. Hence the selection of an animal model with close human anatomical simulation is a desired consideration. Nonhuman primates are the ultimate choice but can be costly. Goats have been widely used for orthopaedic research. Pigs are favoured for cardiothoracic and gastrointestinal research. Rabbits are suitable for ophthalmic surgeries.
- ***Inbred Strain***
  Through brother-sister mating over a minimum of 20 generations, an inbred strain is produced. Animals are virtually identical overtime but less robust than outbred.
- ***Outbred Stock***
  This is bred to minimize inbreeding and ensure genetic variability. They are frequently used in toxicology. As each animal is virtually a unique genetic individual, it is similar to testing a heterogeneous population. The animals have the disadvantage of being phenotypically variable and come with variable genetic make up.
- ***F1 Hybrids***
  A F1 Hybrid is produced when two inbred strains are crossed. Each of the F1 progeny is genetically as uniform as an inbred strain but is more robust. The hybrid vigour is expressed in faster growth, higher survival and longer life span.
- ***Mutant Strains***
  Mutant strains can occur spontaneously or through manipulation in a breeding programme. Mutant strains have their genome changed at least at one locus. One of the most useful strains is the BALB/c athymic ("nude") mouse, which is deficient in T cells (lymphocytes). The deficiency means that the rejection rate to foreign tissues in nude mice is very low.
- ***Transgenic Strains***
  Transgenic animals are produced by the introduction of foreign DNA into the genome. The DNA may direct or block control of physiological parameters such as blood pressure.
- ***Sex***
  The sex of the animal may cause experimental variability. Differences include hormonal cycles, muscle:fat ratio, behavioural differences and variation in growth patterns.

Environmental Factors

The physiological response of an animal is also determined by its dramatype. Russell and Burch used this term to describe the pattern of performance in a single physiological response of short duration relative to the animal's lifetime; for instance, the reaction to a hormone of its target organ, or the reaction of the whole organism to a poison. Variations in such responses are the joint product of two factors. One is the phenotype, the other is the proximate or immediate environment in which the response is elicited. Dramatypic variation thus depends on the animal's more stable properties and is phenotypically determined, and on the environmental conditions in which these are expressed in action.

If we wish fully to control the variance of physiological responses, we must first, control the phenotype, and second, control the environmental conditions in which the animals are tested.

- ***Ventilation***
  There should be good air exchange in the animal room to prevent spread of diseases. Concentration of waste gases must be kept to a minimum. There must be stable temperature and humidity. The adequacy of ventilation in the animal room can often be assessed by the odour level.
- ***Light/Dark Cycle***
  Most animals used in research are nocturnal. The duration, intensity and spectrums of light affect animal biorhythms. Recommendations are that rooms have a 12:12, light/dark cycle using light of 325 lux.
- ***Noise***
  Animals can hear higher frequencies than humans. They may be adversely affected by high frequency machine generated noise. Rats exposed to high frequency noise have elevated corticosterone, and other altered physiological parameters including white cell counts, renal function, blood pressure, blood glucose and estrous irregularities.
- ***Chemical Factors***
  Ammonia is produced from bacteria breakdown of urine and faeces. Ammonia is toxic and levels over 25 ppm can cause pathological changes in the respiratory tract and cornea.
- ***Stress***
  Stress due to transportation, overcrowding, isolation and frequent handling may have physiological consequences. A direct consequence of stress is depressed immune system, which may increase susceptibility to latent oncogenic and infectious agents.

Refinement

Refinement refers to techniques to reduce or eliminate unnecessary pain and distress to study animals. Investigators are required to consider alternatives to painful procedures and to avoid or minimize discomfort, distress or pain, consistent with sound scientific practice and the goals of the research. Some refinement opportunities include:

1. *Pain relieving drugs*
   While it is preferable to design a protocol that avoids pain and distress, when this is not possible, an appropriate plan must be developed for the use of anaesthetics (Appendix 3), analgesics (Appendix 4) or other measures, such as anti inflammatory agents, antibiotics or sedatives. If there is doubt as to whether a procedure will cause pain, a good guide is that if it would cause pain in humans, pain-relieving techniques must be employed.

2. *Non-pharmacological techniques*
   - **New surgical, diagnostic and therapeutic techniques**
     New surgical, diagnostic and therapeutic techniques have the capability to dramatically reduce the invasiveness of procedures for data collection. Examples are the use of sophisticated imaging equipment like ultrasound, CT scan and PET scan and minimally invasive surgical techniques to reduce invasiveness of procedures for data collection. Other examples are advances in technology that allows for analysis of small blood or tissue volume.
   - **Environmental complexity and enrichment**
     Animals used in research are housed in an environment that is very different from their natural habitat. These animals also have social structures, which are inhibited when there are isolated in the laboratory setting. Wherever possible animals must be provided with stimuli that encourage normal behaviour. Environment complexity can be increased by providing apparatus for climbing or sticks for gnawing as appropriate for the species. Judicious use of mirrors may be helpful. Visual and auditory stimuli can be provided by various means e.g. television or cassette player.
   - **Establishment of humane experimental end points**
     Experimental endpoints refer to situations when an animal is withdrawn from the study and treated or euthanized in the interest of animal welfare. The establishment of the earliest possible humane end point consistent with the research design may provide an additional opportunity to significantly reduce pain and distress, thereby refining the experiment. Death as an endpoint is generally unacceptable. Examples of humane experimental end points are: inability to eat/drink, inability to keep upright, inability to ambulate, excessive weight loss (> 20 % of original body weight), tumour exceeds 10 % of body weight.

Prior to embarking on a project involving live animals, due consideration has to be given to its potential merit in the advancement of knowledge and science for human and/or animal welfare against its painful and distressing effects on the experimented animals.

The 3Rs principle is universally accepted as an approach to responsible and ethical use of animals. Replacement refers to techniques to substitute the use of living higher animals for methods using insentient material or species lower on the phylogenetic scale. Reduction refers to the reduction in the numbers of animals used to obtain information of a given amount and precision. Refinement means any decrease in the incidence or severity of procedures applied to those animals that still have to be used. Investigators must at all times regard the animals as sentient and give proper regard to their care and use.

CHAPTER 2.5

USE OF STATISTICS AS DETERMINANT FOR NUMBER OF ANIMALS USED

Huihua Li

One of the "3R's" of biomedical research is to minimize the number of animals used in each experiment if it is impossible to use alternative methods that do not include animals. Minimizing the number of animals used in a project can be achieved through efficient experimental designs, clear understanding of the objectives of the study, controlling variation and appropriate statistical analysis. The sample size of an efficient experiment needs to be large enough to achieve the scientific objectives of the study. However, the sample size should not be unnecessarily large to avoid wasting resources and to minimize distress to animals.

For certain types of experiments, it is impossible to compute the sample size because prior information is lacking or because the success of the experiment is highly variable. For example, in a pilot study, which aims to explore a new research area, the number of animals to be used is based on experience and guesswork as no prior data is available for estimating the number of animals needed for the study. Such exploratory experiments are carried out to generate new hypotheses that can be formally tested. It is, thus, less crucial to formally estimate the sample size, because the aim of the study will be verified by additional experiments. However, most animal experiments aim to verify formal hypotheses, in which case, it is necessary and possible to estimate the number of animals required.

Power Analysis for Determining the Number of Animals to be Used

One of the ways to determine the size of animals used is the "power analysis", which depends on the relationship between six variables (Festing et al., 2002).

- *Effect size of biological interest (δ)*
  The effect size corresponds to how large a biological effect would be of scientific interest. The larger the effect size, the fewer the animals needed to detect it.

- *Standard deviation (σ)*
  The estimated sample size is heavily dependent on the standard deviation. For discrete variables such as survival status (that is dead or live), the standard deviation is a function of the proportion that dies, so there is no need to separately specify it. For continuous characters such as body weight, it is necessary to estimate it. In this case, either similar experiments or reports of similar experiments in the literature or pilot studies can provide an approximate estimate.

- *Significant level (α)*
  This provides a constraint on the probability that the experiment will give a false-positive result. Usually, a significance level of 0.05 is used.

- *Desired power of the experiment ($1-\beta$)*
  The power is the probability of detecting the specified effect at the specified significance level. The power is usually set between 80 % and 90 %. The higher the power, the larger the sample size.

- *Sample size (N)*

- *Alternative hypothesis (that is, a one- or two-sided test)*

Variable

There are three types of variables that an investigator may measure: (1) Binary variable, often expressed as a rate or proportion of a yes/no outcome; (2) Continuous variable, such as the concentration of a substance; (3) Time to event variable, such as the duration before the appearance of disease or death. For different kinds of data, different statistical methods should be used to estimate the number of animals to be used.

1. *Sample size estimation for binary variable (Machin et al., 1997)*
   When the outcome is binary (that is, success/failure), the standard tests for comparing two proportions are either χ^2 test or Fisher's Exact test. The choice of the appropriate test influences the sample size required to detect a difference in proportions. One should use the same test for the planning as for the analysis.

   Usually the data of this type can be summarized in a 2×2 table as follows (Table 2.5.1).

Table 2.5.1: Unpaired 2×2 contingency table

Groups	Success	Failure	Total	Observed proportion of success	Anticipated proportion of success
1	a	c	m	a/m	p_1
2	b	d	$n = \varphi m$	b/n	p_2
Total	r	s	N		

Suppose with equal sample size in each group ($\varphi = 1$), we want to detect an anticipated difference in proportions of δ ($\delta = p_2 - p_1$) at significance level α and power $1 - \beta$ using a two-sided test.

- **Sample size using χ^2 test**
  The required sample size m for Group 1 using the χ^2 test could be given by:

$$m = \frac{\left\{z_{1-\alpha/2}\sqrt{2\bar{p}\left(1-\bar{p}\right)} + z_{1-\beta}\sqrt{p_1\left(1-p_1\right)+p_2\left(1-p_2\right)}\right\}^2}{\delta^2} \qquad (1)$$

where $\bar{p} = \dfrac{p_1 + p_2}{2}$

The total number of sample size required: $N = m + n = 2m$.

If one proportion is known, the required number of animals is given by:

$$N = \frac{\left\{z_{1-\alpha/2}\sqrt{p_1\left(1-p_1\right)} + z_{1-\beta}\sqrt{p_2\left(1-p_2\right)}\right\}^2}{\delta^2} \qquad (2)$$

- ***Sample size using Fisher's Exact test***
  After determining m from Equation 1, the sample size in one group required by Fisher's Exact test to compare two unknown proportions is:

$$m_{Exact} = \frac{m}{4}\left\{1 + \sqrt{1 + \frac{4}{m\delta}}\right\}^2 \qquad (3)$$

The total number of animals required: $N_{Exact} = 2m_{Exact}$

Example 1:
We want to test whether the new drug B is better than drug A by carrying out a study including mice. The pilot study showed that 40 % of the mice treated with drug A responded well while 70 % of the mice treated with drug B responded well. How many mice should be used, using a two-sided χ^2 test or Fisher's Exact test with $\alpha = 0.05$, and power $1 - \beta = 0.80$?

Answer:

$p_1 = 0.4$, $p_2 = 0.7$, $\delta = 0.3$, $z_{1-\alpha/2} = 1.96$, $z_{1-p} = 0.84$.

Assume that equal number of mice are assigned in each group, $\bar{p} = 0.55$.

- The number of mice required in each group using χ^2 test:

$$m = \frac{\left\{z_{1-\alpha/2}\sqrt{2\bar{p}\left(1-\bar{p}\right)} + z_{1-\beta}\sqrt{p_1\left(1-p_1\right)+p_2\left(1-p_2\right)}\right\}^2}{\delta^2}$$

$$= \frac{\left\{1.96\times\sqrt{2\times0.55\times(1-0.55)}+0.84\times\sqrt{0.4\times(1-0.4)+0.7\times(1-0.7)}\right\}^2}{0.3^2}$$

$$= 42$$

Therefore, the number of mice required in each group is 42 and the total number of mice required in this trial is 84 using χ^2 test.

- The number of mice required in each group using Fisher's Exact test:
  Since m = 42, the number of mice required in each group using Fisher's Exact test:

$$m_{Exact} = \frac{m}{4}\left\{1+\sqrt{1+\frac{4}{m\delta}}\right\}^2 = \frac{42}{4}\left\{1+\sqrt{1+\frac{4}{42\times0.3}}\right\}^2 = 49$$

Thus, the number of mice required in this trial using Fisher's Exact test is 98, with 49 mice in each group.

- Free software is available to estimate sample size for comparing two proportions at http://www.biomath.info/power/index.htm.
  Fig 2.5.1 shows the screen at this website. Firstly, we chose the method of chi-square test on proportions under two groups (Fig 2.5.2). Then we used the chi-square test by keying in all the relevant information including proportions in two groups, significance level, power and the sample size ratio of these two groups as shown in Fig 2.5.3. Finally we got the estimated sample size for each group, which is shown in Fig 2.5.4.

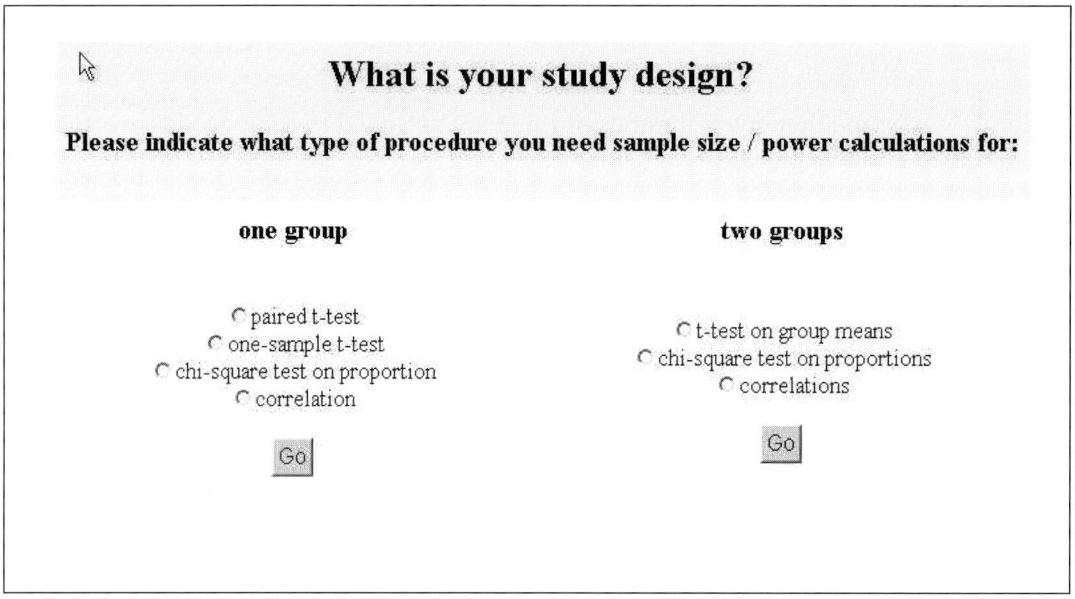

Fig 2.5.1: Screen shown at http://www.biomath.info/power/index.htm.

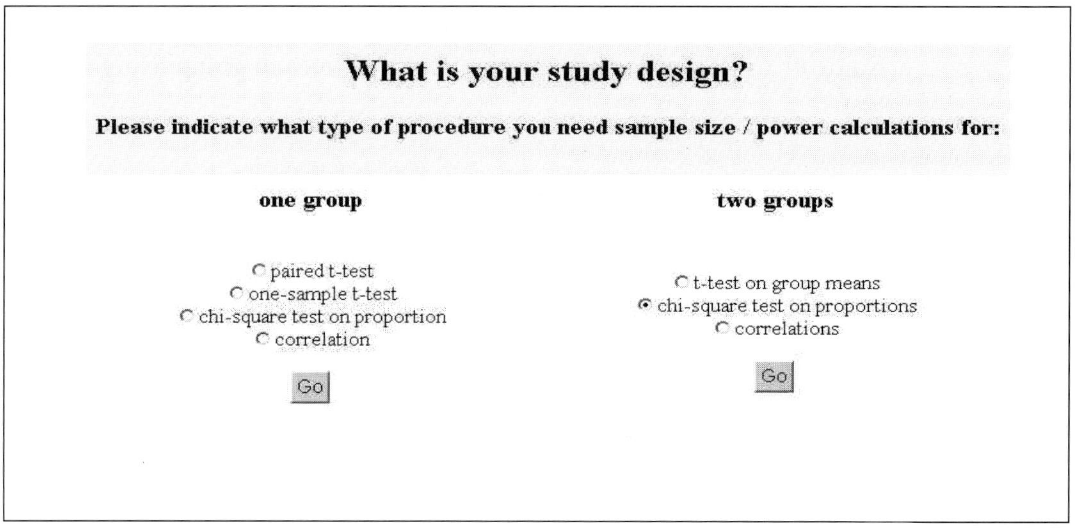

Fig 2.5.2: Screen after selecting the method for comparing two proportions.

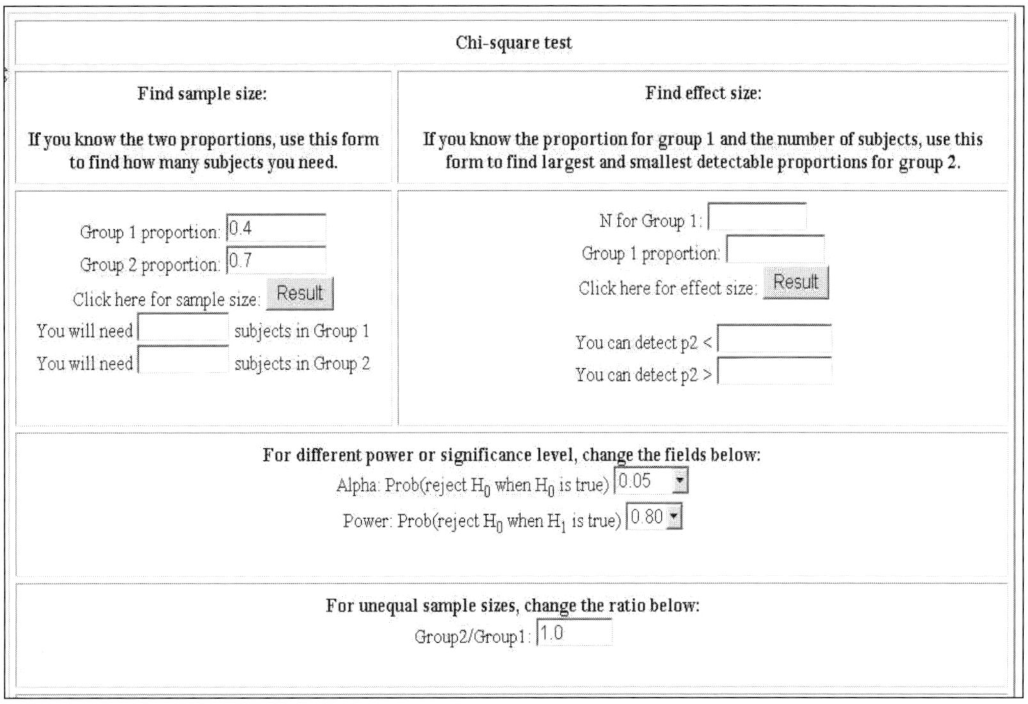

Fig 2.5.3: Screen of sample size estimation for comparing two proportions using Chi-squared test.

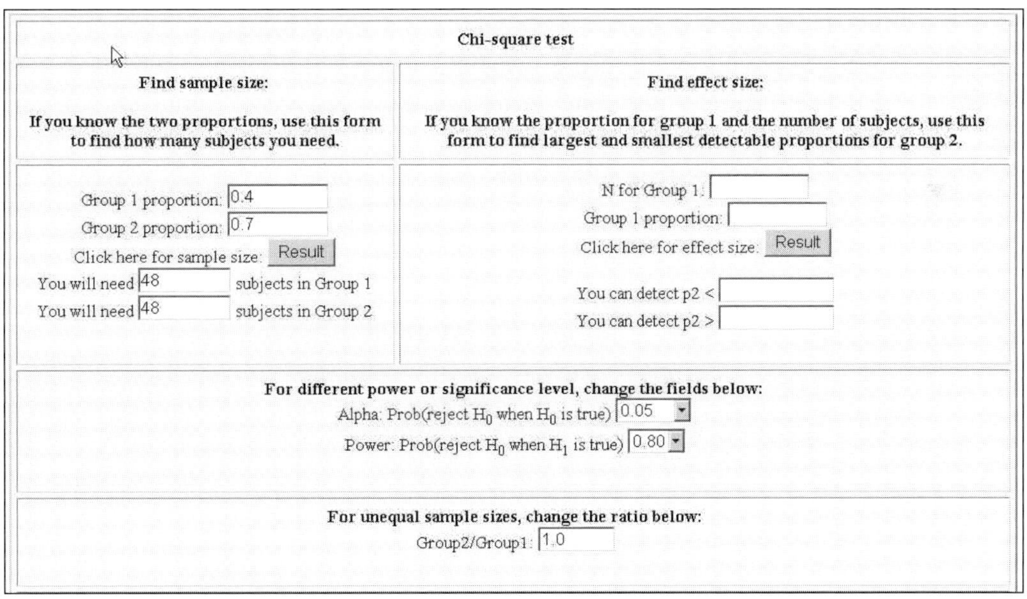

Fig 2.5.4: Screen showing the estimated sample size after running the software.

The number of mice required in this trial using this free software is 96 with 48 mice in each group, which is a little bit different from the one we estimated using Equation 1. Actually there is quite a few equations can be used to estimate sample size; therefore, different software may give you different results.

Example 2:

If in the previous example, we already knew that 40 % of the mice treated with drug A responded well. We still want to test whether the new drug B is better than drug A by carrying out a study including mice. We did a pilot study, which showed that 70 % of the mice treated with drug B responded well. How many mice should be used, using a two-sided χ^2 test with $\alpha = 0.05$, and power $1 - \beta = 0.80$?

Answer:

$p_1 = 0.4$, $p_2 = 0.7$, $\delta = 0.3$, $z_{1-\alpha/2} = 1.96$, $z_{1-p} = 0.84$ and $\delta = 0.3$

- Using Equation 2, the number of mice needed is:

$$N = \frac{\left\{z_{1-\alpha/2}\sqrt{p_1(1-p_1)} + z_{1-\beta}\sqrt{p_2(1-p_2)}\right\}^2}{\delta^2}$$

$$= \frac{\left\{1.96 \times \sqrt{0.4(1-0.4)} + 0.84\sqrt{0.7(1-0.7)}\right\}^2}{0.3^2} = 21$$

Therefore, the total number of mice required is 21.

- Using free software, we went to the website http://www.biomath.info/power/index.htm Choose the method of chi-square test on proportion under one group (Fig 2.5.5). After inputting group proportion, comparison proportion, significant level and power (Fig 2.5.6), the total number of mice required for this study was estimated to be 24, which is shown in Fig 2.5.7.

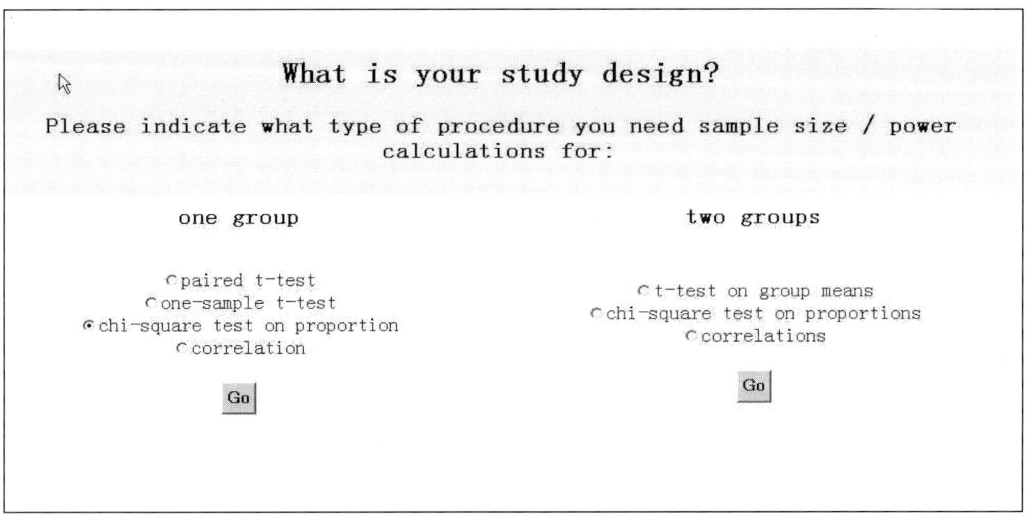

Fig 2.5.5: Screen after selecting the method for comparing one proportion to one known proportion.

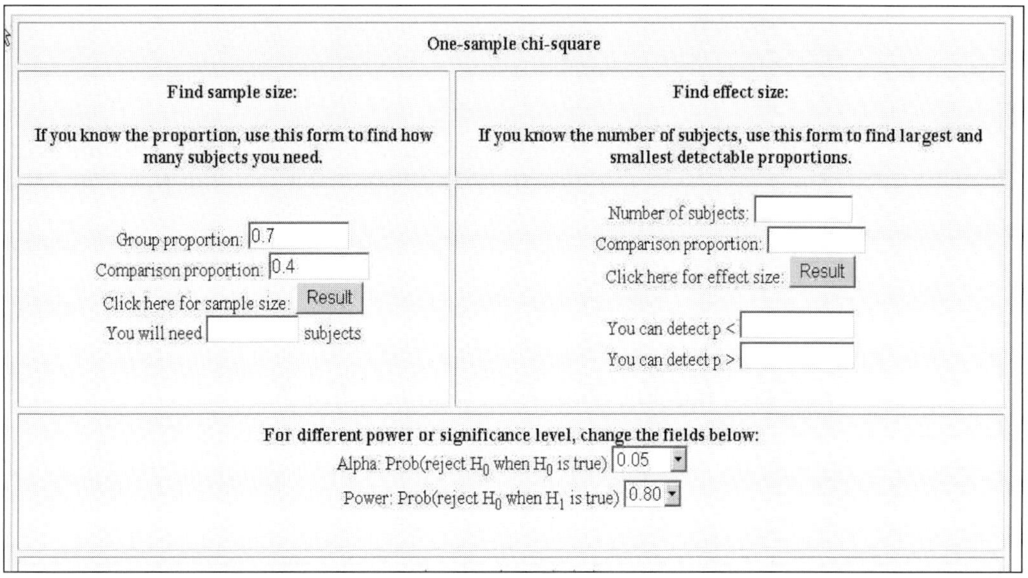

Fig 2.5.6: Screen for comparing one proportion to one known proportion using Chi-square test.

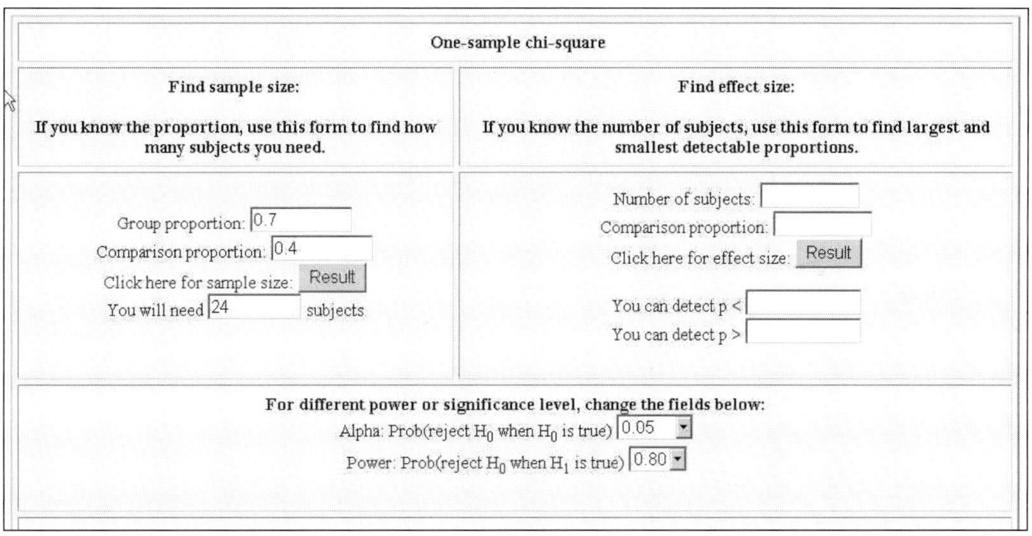

Fig 2.5.7: Screen showing the estimated sample size after running the software.

2. *Sample size estimation for continuous variable*

- **Student's t-test for comparing two means (Machin *et al.*, 1997)**
  If continuous outcome variables are sampled from a Normal distribution, the usual test is the two-sample t-test.
  i. Unpaired Student's t-test
  Given two samples, assume that they have different means, but the same standard deviation α, μ_1 and μ_2 are the two alternative population means, the anticipated (standardized) effect size is:

$$\Delta = \frac{|\mu_2 - \mu_1|}{\sigma}.$$

If the animals are equally assigned to each group, the number of animals required in one group using a two-sided test is given by:

$$m = \frac{2\left(z_{1-\alpha/2} + z_{1-\beta}\right)^2}{\Delta^2} + \frac{z_{1-\alpha/2}^2}{4} \tag{4}$$

The total number of animals required: *N* = *2m*.

If one mean is known, the total number of animals in the group should satisfy:

$$N = \frac{\left(z_{1-\alpha/2} + z_{1-\beta}\right)^2}{\Delta^2} + \frac{z_{1-\alpha/2}^2}{2} \tag{5}$$

Example 3:
The mean body weight of rats used at a certain age is 400 g, with a standard deviation of 23 g. A chemical that changes appetite is to be tested as to whether it alters the body weight of the rats. The scientist would like to be able to detect a 20 g change in body weight between control and treated rats with a power of 80 % at the significance level of 0.05. How many rats should be used?

Answer:

$\sigma = 23$, thus $\Delta = \dfrac{|\mu_2 - \mu_1|}{\sigma} = \dfrac{20}{23} = 0.87$.

- Using Equation 4, the required number of rats in each group is:

$$m = \frac{2\left(z_{1-\alpha/2} + z_{1-\beta}\right)^2}{\Delta^2} + \frac{z_{1-\alpha/2}^2}{4} = \frac{2\times\left(1.96+0.84\right)^2}{0.87^2} + \frac{1.96^2}{4} = 22$$

  Therefore, the total number of rats required is 44, with 22 rats in each group.

- Using free software at http://www.biomath.info/power/index.htm, we chose the method of t-test on group means under two groups (Fig 2.5.8), filled out the form as in Fig 2.5.9. Finally, the number of rats to be used was estimated at 22 in each group (Fig 2.5.10).

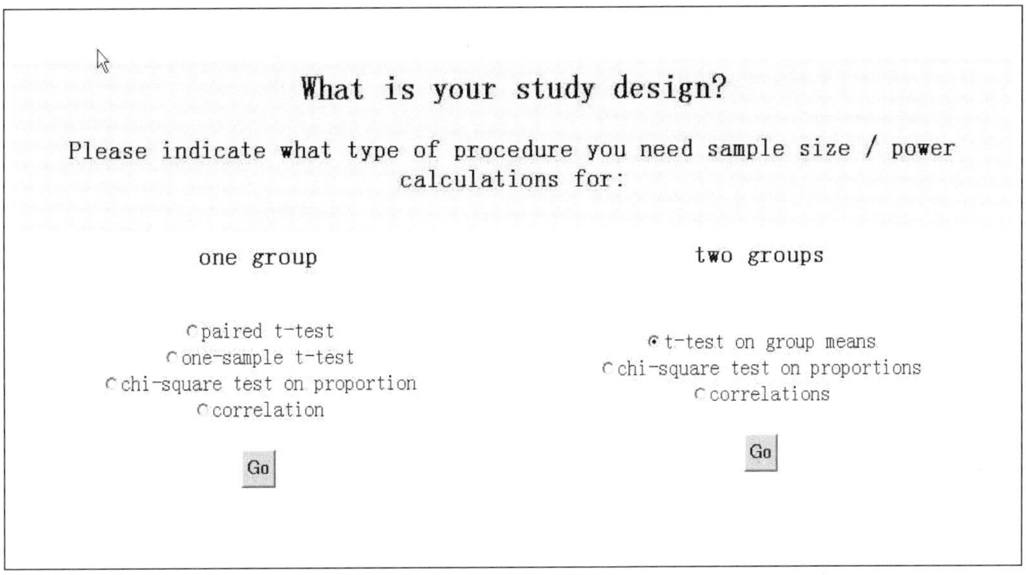

Fig 2.5.8: Screen after selecting the method for comparing two unpaired group means.

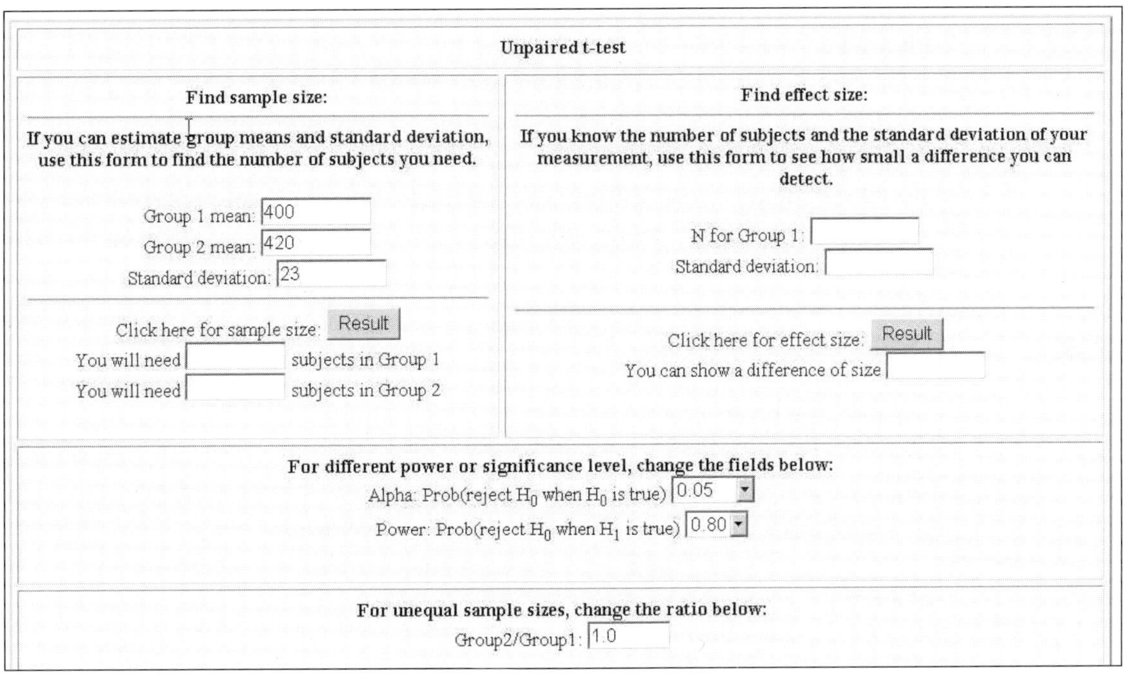

Fig 2.5.9: Screen for comparing two unpaired group means using unpaired t-test.

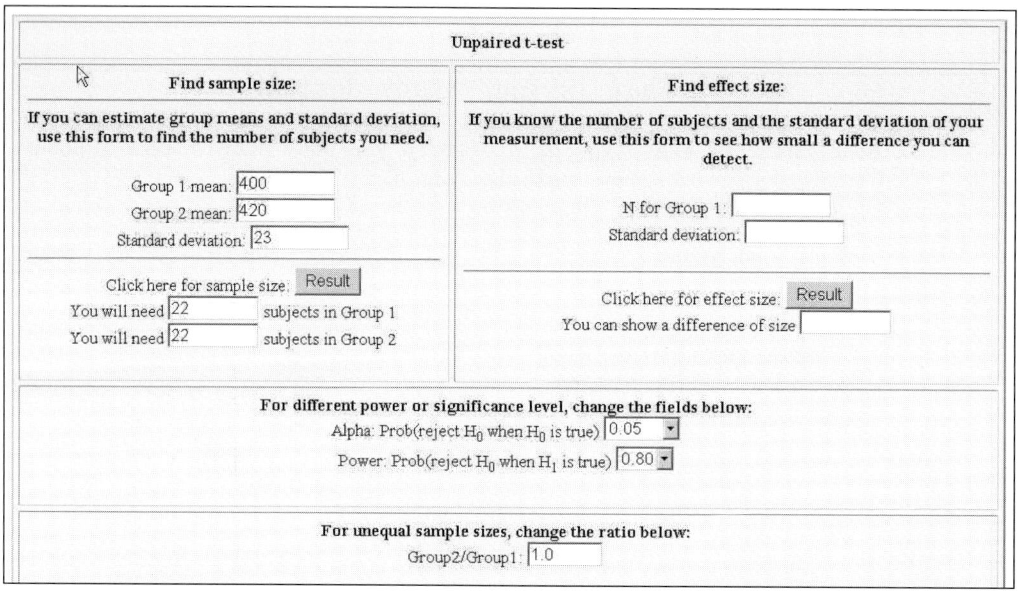

Fig 2.5.10: Screen showing the estimated sample size after running the software.

ii. Paired Student's t-test
Sometimes, the same animal is used as its own control to minimize the variance and reduce the required animals needed to achieve the same power at specified significance level. For example, when one limb of the animal is treated by a new treatment, the other one can be its own control. To detect an anticipated standardized difference Δ at significance level α and power $1-\beta$ using two-sided hypothesis, the required animal size is estimated using Equation 5.

Example 4:
If the study of the previous example is carried out on the same group of rats (measure the body weight before and after the chemical administration for each rat), Paired t-test is used instead of Unpaired t-test. If the standard deviation of the difference is also 23, how many rats should be used at the significance level of 0.05 to achieve the power of 0.8?

Answer:
$\Delta = 0.87$.

- Using Equation 5, the total number of rats required is:

$$N = \frac{\left(z_{1-\alpha/2} + z_{1-\beta}\right)^2}{\Delta^2} + \frac{z_{1-\alpha/2}^2}{2} = \frac{(1.96+0.84)^2}{0.87^2} + \frac{1.96^2}{2} = 13$$

- Using free software at http://www.biomath.info/power/index.htm, we chose the method paired t-test under one group (Fig 2.5.11), filled the parameters as in Fig 2.5.12, the total number of rats needed was estimated at 6-13 (Fig 2.5.13).

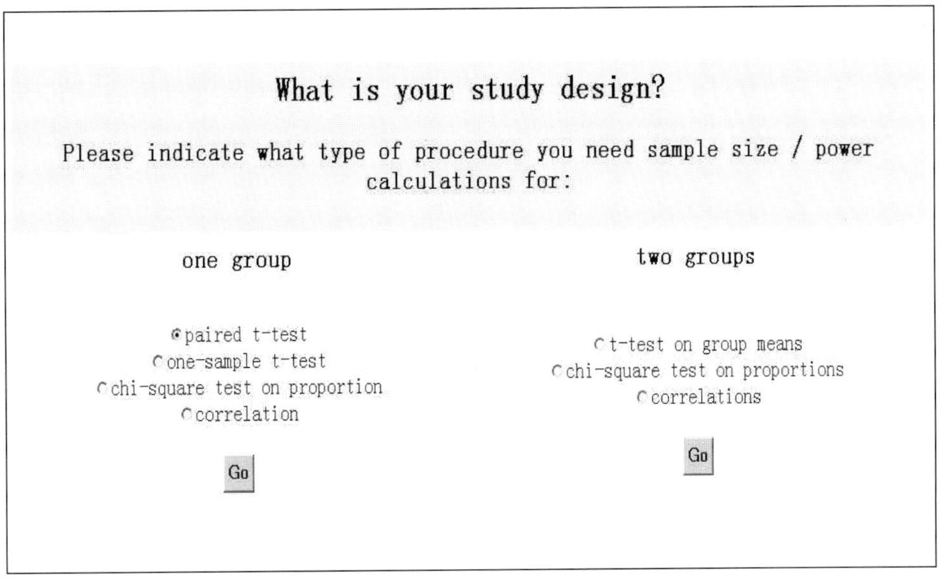

Fig 2.5.11: Screen after selecting methods to compare two paired group means.

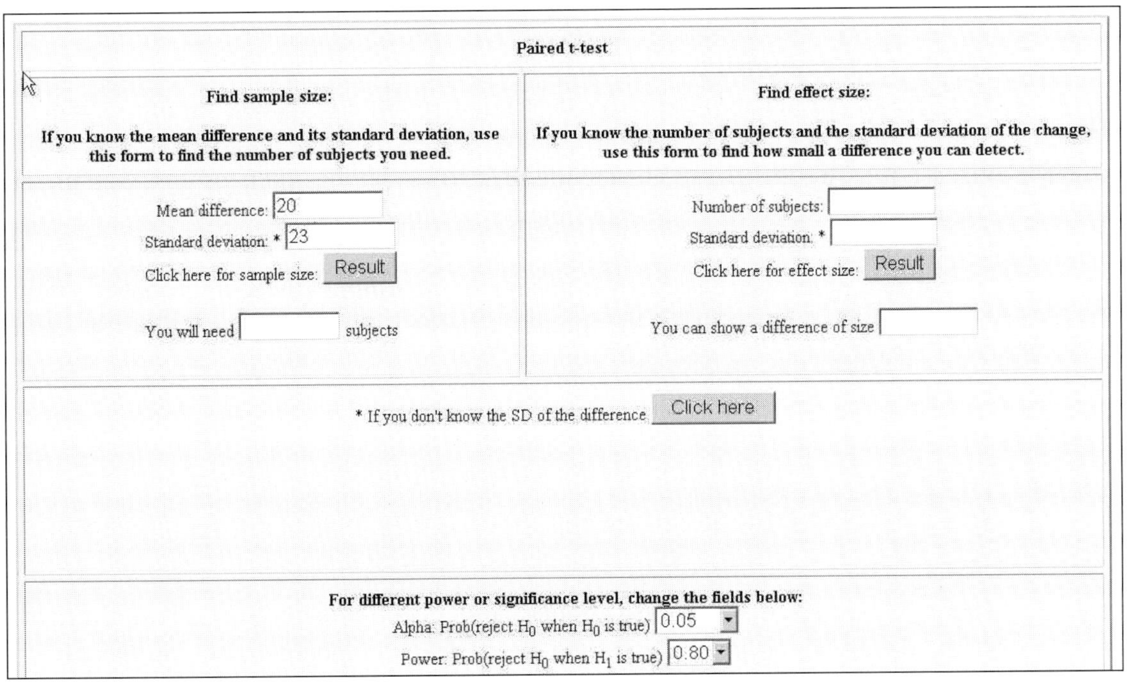

Fig 2.5.12: Screen for comparing two paired group means using paired t-test.

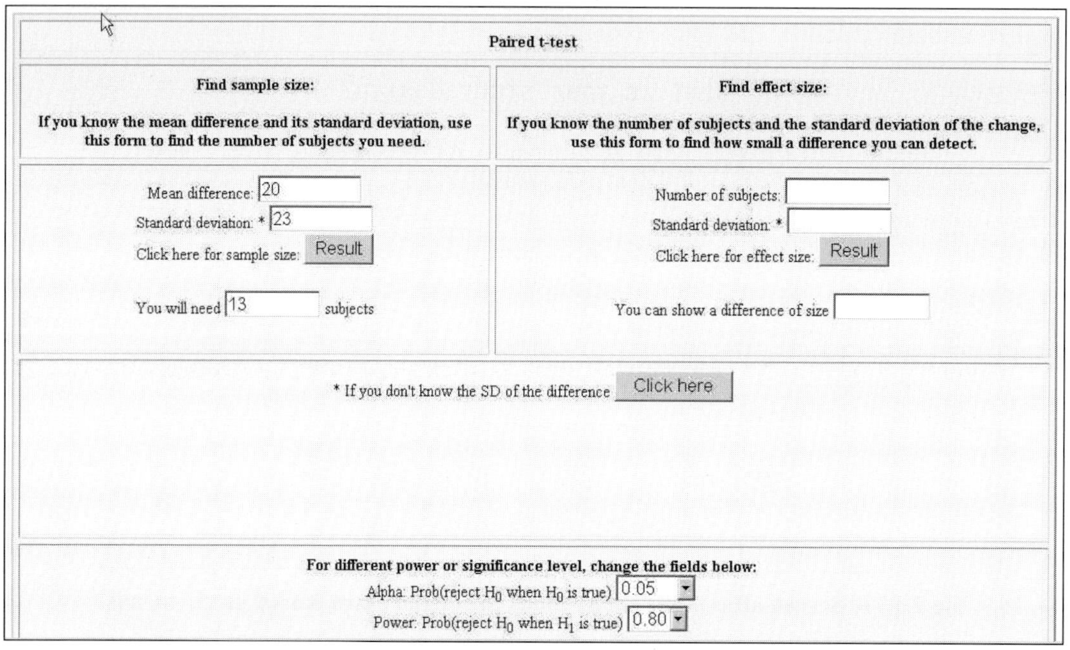

Fig 2.5.13: Screen showing the estimated sample size after running the software.

- **ANOVA for comparing more than two means**

Analysis of variance (ANOVA) is used to compare means of more than two groups, when the variances within each group are approximately equal and the observations in one group are independent of those in other groups. Sample size calculation using ANOVA is possible by several ways. Here one possible way is described as follows (Chow, 1998).

For example, we may have k groups of treatment (including control) with n subjects in each group, with the intention being to investigate whether the effects of these treatments are the same or not.

Let

$$\delta^2 = \frac{n\sum_{i=1}^{k}\tau_i^2}{\sigma^2},$$

where σ^2 is the within group variance, and τ_i is the difference of the average of all the observations in k groups and the average of observations in group i. The required sample size then can be determined by solving the following equation:

$$z(\beta) = \frac{\sqrt{v_2\left[2\left(v_1+\delta^2\right)^2-\left(v_1+2\delta^2\right)\right]}-\sqrt{v_1\left(v_1+\delta^2\right)\left(2v_2-1\right)F_A}}{\sqrt{v_1\left(v_1+\delta^2\right)F_A+v_2\left(v_1+2\delta^2\right)}} \qquad (6)$$

where $F_A = F(\alpha, v_1, v_2)$; $v_1 = k-1$; $v_2 = k\left(n-1\right)$.

If the power is set at 0.8, to achieve this power, the required sample size used should make $z(\beta)$ at least 0.85.

Example 5:
Suppose we want to evaluate the effect of 3 treatments for a particular cancer by comparing the corresponding reduced tumour volume in mice given each treatment. Assume the tumour volume reduced by 10 units for treatment A, by 8 units for treatment B and by 6 units for treatment C, with the within group variance being 9 for all these treatments. How many mice should be used at significance level of 0.05 and power of 0.80?

Answer:
$\mu_1 = 10$, $\mu_2 = 8$, $\mu_3 = 6$ and $\sigma_2 = 9$. With equal number of mice n in each group, $\overline{\mu} = 8$.

Thus $$\delta^2 = \frac{n\sum_{i=1}^{k}\tau_i^2}{\sigma^2} = \frac{n\sum_{i=1}^{3}\left(\mu_i - \overline{\mu}\right)^2}{\sigma^2} = \frac{n\left\{(10-8)^2 + (8-8)^2 + (6-8)^2\right\}}{9} = \frac{8n}{9}.$$

If $n = 9$, $\delta^2 = 8$, $v^1 = 2$, $v^2 = 24$ and $F_A = 3.40$.

In this case:

$$z(\beta) = \frac{\sqrt{v_2\left[2\left(v_1+\delta^2\right)^2 - \left(v_1+2\delta^2\right)\right]} - \sqrt{v_1\left(v_1+\delta^2\right)\left(2v_2-1\right)F_A}}{\sqrt{v_1\left(v_1+\delta^2\right)F_A + v_2\left(v_1+2\delta^2\right)}}$$

$$= \frac{\sqrt{24\left[2\times(2+8)^2 - (2+2\times 8)\right]} - \sqrt{2\times(2+8)(2\times 24-1)\times 3.40}}{\sqrt{2\times(2+8)\times 3.40 + 24\times(2+2\times 8)}}$$

$$= 0.43$$

In order to achieve the power of 0.8, $z(\beta)$ should be at least 0.85. Thus, 9 mice in each group are not enough. So we iterate this process by increasing the number of mice used to 12 mice in each group with $z(\beta)$ of 0.87. Therefore, the total number of mice required is 36 with 12 mice in each group.

3. *Sample size estimation for time to event variable*
Although the statistical analysis of time to event variable is complicated, simple approaches can be adopted to estimate sample size for this type of variable.

- One possible way is to estimate sample size using the proportions in the two experimental groups exhibiting the event by a certain time, which converts time to an event into a binary variable. Sample size can be estimated by Equation 1 or by means of the method of comparing two proportions at http://www.biomath.info/power/index.htm.

- The other possible way is to treat time to an event variable as a continuous variable. However, this approach is applicable only if all animals are followed to event occurrence (for example, until death or time to exhibit a disease). Sample size can be estimated using Equation 4 or the method of comparing two unpaired group means at http://www.biomath.info/power/index.htm.

CHAPTER 2.6

THE ADVANTAGES OF ACCREDITATION WITH AAALAC

Bryan Ogden

Accreditation with the Association for Assessment and Accreditation of Laboratory Animal Care International (AAALAC) is a major achievement of great value to research institutions. AAALAC is a private, nonprofit organization that promotes the responsible treatment of animals in science through a voluntary assessment and accreditation programme. Accreditation by AAALAC indicates an institutional commitment to maintain a quality animal care and use programme. Application of AAALAC standards ensures high-quality research and animal care, resulting in better science. Resources and services provided by AAALAC help raise the level of animal care world-wide. Institutions with AAALAC accreditation recognize ongoing benefits through participation in the programme.

History and Organization

During the last 50 years, society has witnessed the emergence of new fields of science and countless advances in technology and medicine never before imagined. Virtually every major medical breakthrough credits research involving animals. Leading veterinarians and researchers saw the need for an independent organization to encourage high standards for humane animal care and use in science. AAALAC was created in the United States to meet this need, and was incorporated as an independent nonprofit organization in 1965. Founders included 14 charter members representing national professional scientific, medical, and educational organizations Today more than 65 "Member Organizations" govern AAALAC, and are represented on the Board of Trustees. The 40-member "Council on Accreditation" carries out the international accreditation programme. The Council is comprised of North American and European Sections. There are more than 180 "ad hoc Consultants" who help conduct programme evaluations. More than 700 companies, universities, hospitals, government agencies and other research institutions in 29 countries have earned AAALAC accreditation, demonstrating their commitment to responsible animal care and use.

Standards

AAALAC is non-regulatory and does not formulate regulations. Rather, the programme demonstrates conformance with accepted practices, guidelines and Federal, State, and local regulations. These become the standards AAALAC uses to evaluate and assess compliance. The most well-known guidelines are those in the *Guide for the Care and Use of Laboratory Animals*, published by the National Research Council (USA), and most recently revised in 1996. The *Guide* is the primary resource used by AAALAC's Council on Accreditation to evaluate animal programmes, and is widely recognized throughout the international scientific community. Other widely accepted guidelines include the American Veterinary Medical Association's "Report of the AVMA Panel on Euthanasia," and the American Journal of Veterinary Research's "Guidelines for Animal Surgery in Research and Teaching." Copies of these publications and a list of additional resources and guidelines are available through AAALAC by calling 1.301.696.9626, or visiting the "Resources" section of AAALAC's Web site (www.aaalac.org).

Most countries have their own regulations and standards for animal care and use in science. Because there has been inconsistency among country standards, the international scientific and animal welfare communities continue to work toward generating legislation and guidelines that harmonize regulations across borders. For example, the Council for International Organizations of Medical Sciences (CIOMS), an international nongovernmental organization, published the "International Guiding Principles for Biomedical Research Involving Animals" in 1985, which has provided basic guidelines for many countries. AAALAC has copies of these documents, along with some specific country information. Each time a new institution or company becomes accredited, it helps to raise the *global* benchmark for animal well-being in science.

Animal research programmes must comply with applicable legislation, regulations, policies and guidelines in their country. In addition, many top institutions *voluntarily* choose to go beyond the minimums required by seeking accreditation through AAALAC International. In Singapore the National Advisory Committee for Laboratory Animal Research (NACLAR) has established guidelines that are very similar to the US G*uide*. The Animals and Birds Act in Singapore provides the legal basis for inspection and licensing of animal research facilities by the Agri-food and Veterinary Authority (AVA). The AVA uses the NACLAR Guidelines as the standard for evaluations. As of July 2006 two Singapore research institutions have been granted AAALAC Accreditation, Singapore General Hospital Department of Experimental Surgery and Maccine Pte. Ltd.

Becoming AAALAC Accredited

The accreditation process is dynamic, requiring institutions to conduct their own extensive internal review, as well as undergo a comprehensive, on-site assessment by AAALAC evaluators. It is meant to be educational, and includes an extensive self-evaluation of the animal programme during which a Programme Description is written using an outline provided by AAALAC. The Programme Description covers all aspects of support for animal care and use, from facility and housing for animals to all programmatic issues concerning animals including management, care, and use. These are divided into four major categories as outlined in the *Guide:* institutional policies, laboratory animal management, veterinary care, and physical plant.

The Programme Description, including all aspects of animal care and use at the institution, is submitted to AAALAC along with an application for AAALAC accreditation. Next, an AAALAC team visits the facility. The site visit team is comprised of at least one member of AAALAC's Council on Accreditation and one or more AAALAC *ad hoc* consultants, many of whom are bench scientists. During their review, the team assesses the programme to verify that it is upholding the principles outlined in the *Guide for the Care and Use of Laboratory Animals (NACLAR Guidelines* in Singapore*),* local regulations and other appropriate reference resources. Professional judgment and performance-based criteria are incorporated into the evaluation. This is a peer-review process with open discussion. Open discussion, expert to expert, facilitates a two-way learning process.

Site visitors walk though the facility and review documentation. They view the animal environment, housing and management. They review the health of the animals and the quality of veterinary medical care, including preventive medicine, surgery, pain management, euthanasia, procurement and transportation. They assess the adequacy of the physical plant, including functional areas, surgery facilities, and building management. They review the function of the Institutional Animal Care and Use Committee (IACUC), including animal use protocols, the protocol approval process, meeting minutes, programme evaluation and facility inspection reports, and compliance monitoring. They review animal care and use policies and responsibilities, including standard operation procedures (SOPs). Programmes for training of personnel, occupational health and safety, and disaster planning are evaluated along with relevant documentation. They pay particular attention to signs of institutional support for the animal care and use programme.

At the conclusion of the site visit, an exit briefing is held by the site visitors to discuss preliminary observations and to answer questions raised during the visit. This step has been viewed as instructive for both the programme staff and the site visitors. The institution is allowed two weeks to respond to the findings outlined in the exit briefing. The team submits a report, which includes commendations and recommendations, to AAALAC's Council on Accreditation (Council). The report is then reviewed and deliberated on by Council members and the accreditation status is determined. If deficiencies are found, they are outlined in a letter and the institution is given a period of time to address them. After the deficiencies are corrected, accreditation is awarded. This entire process is completely confidential, allowing frank and open dialogue between them and the institution and AAALAC International.

Once accreditation has been awarded it must be maintained. Institutions are required to submit annual reports to AAALAC. Every three years the institution must submit an updated Programme Description and schedule another site visit. The site visit process, reporting, and review by AAALAC Council are repeated. Those institutions that continue to maintain a quality animal care and use programme are awarded Continued Full Accreditation. If an institution is found to have deficiencies, AAALAC will work with them to help correct those deficiencies.

The AAALAC accreditation programme is available to any active laboratory animal programme that uses and cares for animals in research, teaching, or testing. The size of a programme is not an obstacle in attaining accreditation. Humane animal care and use standards are the same for small and large programmes. Concerns about deficiencies should not be a deterrent to seeking accreditation. Awareness of deficiencies and knowledge of their implications are paramount to maintaining high standards of laboratory animal care and use. A detailed description of the AAALAC process is posted at ***www.aaalac.org/ accreditation/index.cfm***.

Programme Status Evaluations

For institutions that are unsure about whether or not they are ready for accreditation, AAALAC offers an assessment service. The decision to offer assessment services (in addition to the accreditation programme) was prompted by a number of requests from non-accredited institutions for a "pre-AAALAC" site visit. These institutions, particularly those outside of the United States, are typically less familiar with the accreditation process and want to find out how their programmes compare to AAALAC standards—before they participate in the formal accreditation programme. AAALAC accommodates these requests through its "Programme Status Evaluation" (PSE) service.

The objective of the PSE service is twofold. First, it's meant to assist institutions in determining if their animal care and use programmes meet AAALAC standards by identifying weaknesses and suggesting ways to improve or correct them. Second, it's meant to familiarize institutions with the AAALAC accreditation process and encourage them to participate.

Entitlements of AAALAC Accreditation

Institutions accredited with AAALAC are entitled to the following:

- Receive AAALAC's proprietary electronic newsletter, the "AAALAC E-brief."
- A free subscription to AAALAC's *Connection* newsletter.
- Telephone and e-mail consultations.
- Listing in the online "AAALAC Directory of Accredited Programmes."
- Access to AAALAC's Members' Only Website:
    - Materials to promote the institutions accreditation
    - "Keeping Connected" (a compilation of news articles and meeting announcements of interest to the animal care community)
    - Tools for maintaining accreditation.
- Periodic updates on the accreditation programme and animal care and use issues.

Quality Animal Care, Better Science

AAALAC International enhances life sciences by promoting the responsible treatment of animals used in research, teaching and testing. For some, animal research is a controversial topic. But like others in the animal welfare arena, AAALAC endorses the use of animals to advance medicine and science when there are no non-animal alternatives, and when it is done in an ethical and humane way. When animals are used, AAALAC works with institutions and researchers to serve as a bridge between progress and animal well-being. This is done through AAALAC's voluntary accreditation process in which research programmes demonstrate that they meet the minimum standards required by law, and are *also* going the extra step to achieve excellence in animal care and use. In this way, **AALAC** International is where Science and Responsible Animal Care Connect.

AAALAC accreditation is a distinction that is widely recognized throughout the international scientific community as assurance of responsible and ethical animal care and use, *and* quality science. AAALAC-accredited organizations should be proud that they are doing their part to raise the global benchmark for animal well-being in science.

Ongoing Benefits

There are numerous benefits received by AAALAC accredited institutions and the animals under their care. Experimental variables are minimized, ensuring high quality reproducible data and scientific validity. A high level of animal husbandry and care and humane animal use are ensured. A high level of occupational health and safety is assured. Quality programmes help recruit quality people. AAALAC accredited institutions demonstrate accountability, showing a real commitment to humane animal use. This assures funding sources, potential clients, and the public that they go beyond minimal requirements in animal use and care. When controversies surrounding animal use in research arise, AAALAC-accredited organizations are consistently better able to withstand the public scrutiny of their research programmes, to justify the need for animal research, and to demonstrate their accountability to the public.

The following are comments on what people value most about AAALAC accreditation:

- "It assures the credibility of our programme among funding sources."
- "It provides the public with a positive image."
- "It helps convince management of the need to commit resources to maintain a high-quality programme."
- "It conveys a high level of professionalism to the scientific community."
- "Application of AAALAC standards ensures high-quality research and animal care, resulting in better science."
- "Completing the Programme Description helps us identify weaknesses and self-correct them."
- "It assures our customers that we have a quality programme."
- "The rigorous peer-review evaluation ensures that we're doing things right."

Conclusion

As the issues and needs of animal research have changed, AAALAC has attempted to adapt and evolve as well. The accreditation programme has maintained consistency in evaluating participants, thus asserting its continued value. The life sciences community has learned the necessity of accountability in addressing issues, especially where the public is concerned. The animal user community has the opportunity to demonstrate beyond a reasonable doubt that they can be held accountable for the humane care and use of animals, and have established effective mechanisms to monitor their activities. Attainment of accreditation attests to the fact that a programme has considered essential elements of animal care and use and abides by them. Those programmes accredited by AAALAC take pride in their achievement. AAALAC invites all inquiries about accreditation and will work with any animal programme interested in achieving accreditation.

AAALAC Contact Information:

In Asia
Pacific Rim Office
AAALAC International
68-3549 Makana Aloha Pl.
Waikoloa, HI 96738
tel: 808.883.2186
fax: 808.883.1155
email: pacificrim@aaalac.org

In North America, South America:
AAALAC International
11300 Rockville Pike, Suite 1211
Rockville, Maryland 20852 USA
tel: 301.231.5353
email: accredit@aaalac.org

In Europe:
AAALAC International
Avenue de Tervuren 402
1150 Brussels Belgium
tel: +32.2.761.6678
email: accredit_europe@aaalac.org

CHAPTER 3

ANIMAL HANDLING AND SURGICAL PROCEDURES

CHAPTER 3.1

GENERAL HANDLING, RESTRAINT, ORAL DOSING/GAVAGE AND INJECTIONS IN LABORATORY ANIMALS

Bryan Ogden

Some of the most common, approved methods for handling, restraint, oral dosing/gavage, and injections in mice, rats, rabbits, hamsters, guinea pigs, nonhuman primates, pigs, and small ruminants (goats and sheep) are described below, with emphasis on mice and rats. Alternative methods may be acceptable, but it is recommended that proposals to use alternatives be approved by the veterinary or technical staff within the animal unit. The most humane methods should always be used, as proper techniques will minimize risk of animal bites and scratches to personnel or unnecessary duress or injuries to the animals.

Handling and Restraint

Although there are significant species differences when handling and restraining an animal, there are several important concepts that apply generally to all species. These include:

1. Handle animals gently, but firmly.
2. Approach an animal slowly, but purposefully.
3. Wear disposable gloves whenever possible.
4. Always wash your hands after removing gloves.
5. Wear a clean laboratory coat and other PPE for each species as odours of other species or blood may be distressing to animals and you can act as a means of spreading infectious agents from one group of animals to another.
6. Use a method appropriate to the species and study.

Animals to be used in experimental protocols that involve extensive manipulation should be handled frequently before the onset of the study. This allows the animals to habituate to the method of being handled and your scent making them more docile while restrained. Within species, particular stocks or strains of animal may have distinctive behavioural responses that may impact the method of restraint.

1. *Mice and Rats*

   Adult mice and rats are picked up by grasping the base of the tail between the thumb and forefinger and gently placing them onto a solid surface. Tails can be injured if grasped near the tip. If transporting a mouse or rat short distances within a room the animal can be rested on the dorsal aspect of your forearm while still grasping the tail. These animals should not be carried by their tails for more than a few seconds! Alternatively, the animal can be placed in a small container with a cover containing holes that admit in air, but do not allow for escape. Mice can also be picked up using a pair of forceps (toothless/atraumatic) to grasp the mouse by either the tail or the skin over the shoulders. However, this method is not acceptable for adult rats.

   Very young rodents (pups), less than 10 days old, can be picked up by cupping the hands around the whole body. This method can be used for picking up a litter (group) of pups, but it is recommended that a small amount of nesting material or bedding be picked up with them.

   For restraint, the mouse is picked up by the tail as described above and placed over the wire bar lid of the cage and lowered until the mouse grasps the wire with its forefeet (Fig 3.1.1). The loose skin over the animal's neck and shoulders is grasped between your thumb and forefinger (index finger). The hand is rotated so that the mouse is lying on its back within the palm of the hand. The animal's head is closest to your thumb while the tail is pressed against the palm with your smallest finger. The result is a mouse that is immobilized for examination or manipulations, such as gavage or injections (Fig 3.1.2).

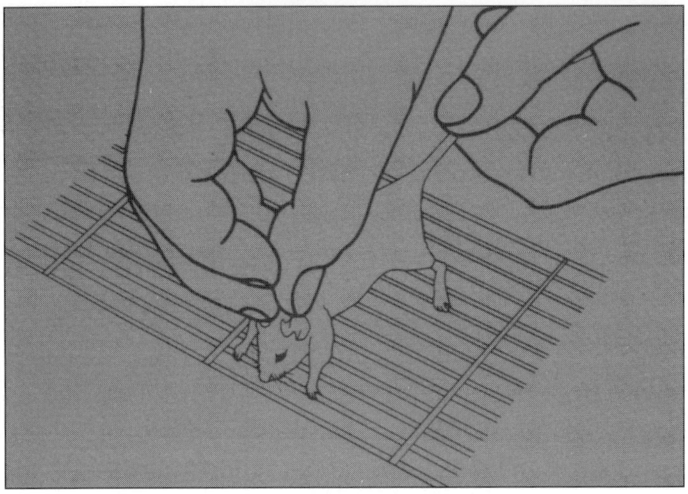

Fig 3.1.1: Mouse is placed over the wire bar lid for the mouse to grasp the lid using its forefeet.

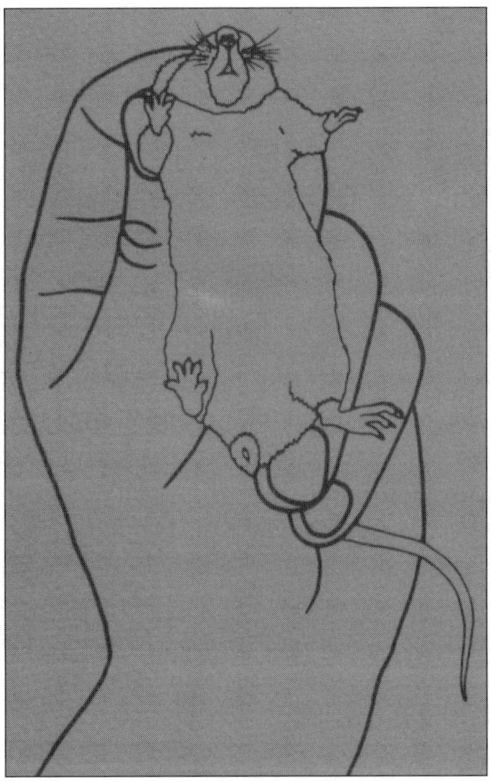

Fig 3.1.2: Handling of mouse.

In preparation for examination or manipulations, the rat may be lifted by the tail and placed on your opposite forearm (a sleeved gown or laboratory coat is essential to protect the arm). The grasp of the tail should then be transferred to the hand of that forearm between the thumb and palm. The opposite hand is used to grasp the animal's body from above (dorsal surface) using one of the four methods. The first method is performed by grasping the rat using the index and middle finger placed firmly on either side of the neck or an alternative method as shown in Fig 3.1.3. The thumb and ring finger are placed on either side of the thorax behind the front legs.

The second method is performed by using the entire hand to grasp the rat firmly around the thorax with your thumb and forefinger placed on either side of the animals' head at the level of the mandible. The third method is performed by grasping with your thumb and middle finger on either side of the thorax behind the front legs and the index finger placed on top of the head between the ears or an alternative method as shown in Fig 3.1.4. The head is then pushed ventrally and the front legs are pushed cranially. This traps the head between the front legs. When held firmly, the rat is restrained and is unable to move its head to bite. With the first three methods the hindquarters should be supported with the other hand or against your chest. Care must be taken with all methods to avoid compressing the chest, since this may inhibit respiration and cause the rat to panic. The fourth method involves grasping loose skin along the rat's back from the neck to the lumbar area while the rat is on the handler's forearm or on a smooth surface.

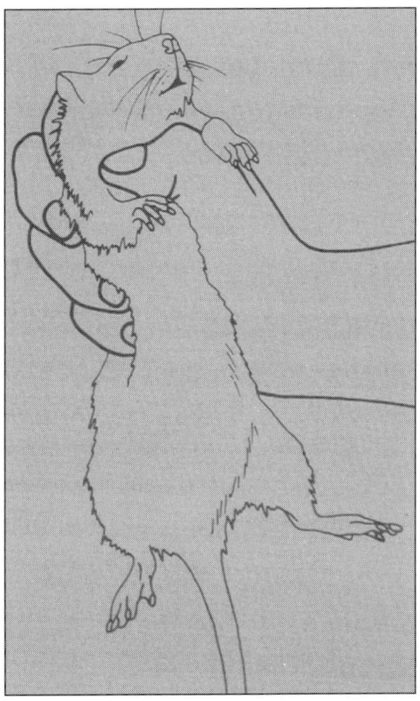

Fig 3.1.3: An alternative to the first method of handling rats.

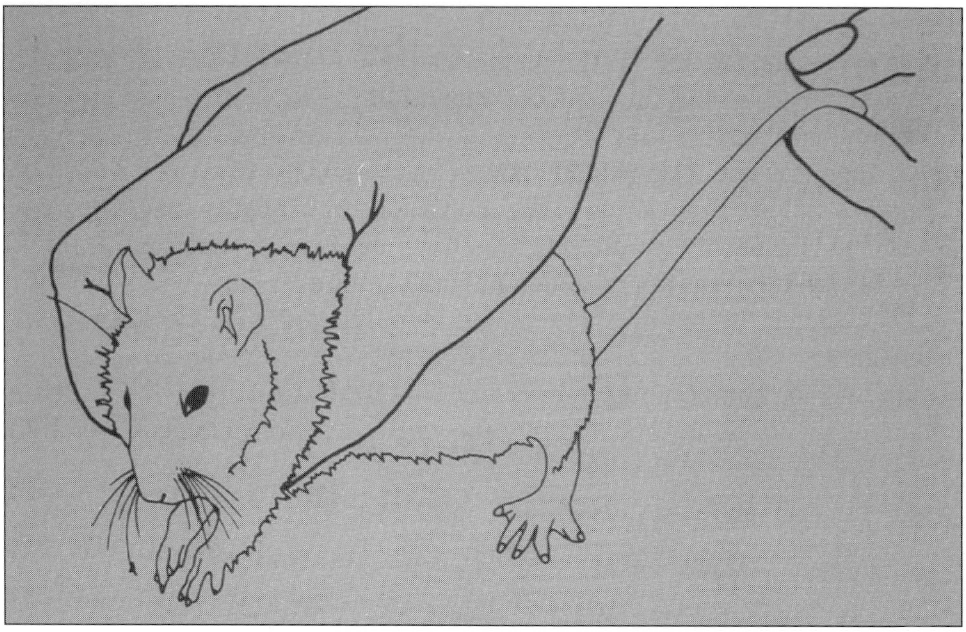

Fig 3.1.4: An alternative to the third method of handling rats.

Another method for restraining rats can be called the "gun-in-holster" technique since the animal is held next to the handler's body near where an American western gunslinger might carry his gun on his gunbelt at his hip. This is done by lifting the rat by the tail and quickly placing the palm of your opposite hand over the dorsum of the rat on your hip. The palm restrains the animal from behind the ears to the pelvis and prevents flexion of the rat's back. The rat's pelvis is restrained between the thumb and index finger. This is done by moving the thumb of the restraint hand anterior to the nearest rear leg and extending the thumb to the rat's pubis. The index finger of that hand is moved behind that rear leg and extended to the caudal-most part of the pelvis. Once the restraint is secured, the tail can be released by the opposite hand. Using this restraint method, injections can be given in the anterior or caudal thigh muscles. The ventral abdomen can be exposed for intraperitoneal (IP) injections by rotating the hindquarters with the restraint hand. It is important that the rat be positioned in a vertical position with the head pointed towards the floor to allow some of the major abdominal organs to fall away from the IP injection site.

Devices are available to restrain mice and rats for a variety of procedures. Commercially available plexiglass restraining cylinders provide access to the animal's tail for intravenous injection or blood collection. Homemade devices for mice can be made out of plastic syringe casings or centrifuge tubes, with breathing holes. Rats can also be restrained by placing them on a solid surface and putting a fabric towel over the head and thorax. One hand is used to grasp the covered portion of the rat. This allows access to the tail and hindquarters.

Conical plastic sleeves, referred to as Decapicones®, can also be used. The plastic is approximately the same thickness as that of heavy-duty plastic waste bags. The flexible, transparent-plastic sleeve is conical, is open at its base, and has a small breathing hole at the apex. The mouse or rat is slid into the cone through the base with its nose resting adjacent to the breathing hole. The excess plastic is gathered and a rubber band is placed around the base of the animal's tail and the plastic of the cone. The cone permits access to the tail and also, if the animal is positioned properly, will permit injections through the plastic wall.

Alternatively, the wire bar lid from a shoebox cage that contains a food trough can be used to restrain a mouse to provide access to its tail. The wire bar lid is set on a solid surface so that it rests on the angular food trough. The mouse is directed between the food trough and the end of the wire bar lid, which is resting on the top's surface. The mouse's tail is directed between the wire bars and gently pulled so that the animal's rear end is held firmly against the lid. This method provides access to the tail, while limiting the mouse's ability to turn around and bite.

2. *Rabbits*
Rabbits can be handled and lifted by grasping the loose skin over the neck and shoulders (the scruff). When lifting by the scruff the hindquarters must be supported with the other hand to keep the back flexed and to prevent struggling (Fig 3.1.5). It is essential that the animal not be allowed to kick backward as it is lifted since this may cause fractures or luxations to the spine. Another reason rabbits should not be allowed to kick backwards is for the personnel safety, since the rear claws of rabbits can cause deep scratches on the handler. Never pick up or restrain a rabbit by the ears since this inflicts needless pain and causes the animal to struggle violently to free itself, even to the point of breaking its neck. A rabbit can be carried in your arms with the head buried between your upper arm and thorax and your hand on the hindquarters to maintain the back in flexion. It is advisable to retain your hold on the scruff with your opposite hand, unless the rabbit is very docile. If a rabbit begins to struggle while in your arms it may be necessary to drop to one knee and include your upper leg in the restraint. In some cases the solution may be to quickly, but gently place the struggling rabbit on another surface such as a table or the floor. When returning a rabbit to a cage, place its rear quarters in first with the head facing away from the cage. This reduces the chances that the rabbit will see the cage and struggle to free itself from restraint in order to jump into the cage.

To examine the abdomen or perineum the rabbit should be placed on its back while holding the scruff. A rabbit may be made to enter a torpid state if placed and held on its back for a few seconds.

Restraint methods for manipulations include hand restraint, wrapping the rabbit in a towel or using a commercially available restraint box or cat bag. Hand restraint for injections usually involves holding the rabbit on a smooth surface, such as a table or countertop. The scruff is held for subcutaneous injection. For IM injections one forearm is used to press the rabbit against your body with the rabbit's head near your elbow and your hand cupped behind the tail. For IP injections (not generally recommended) the rabbit's rear legs are held in one hand and the forequarters are held between your knees in a position to expose the rabbit's ventral abdomen. Commercially available restraint devices usually rely on limiting movement by encircling the rabbit's neck and holding the body in place. Care must still be taken to prevent struggling and it may be necessary to place a folded towel behind the rabbit while in a restraint container to keep the back flexed.

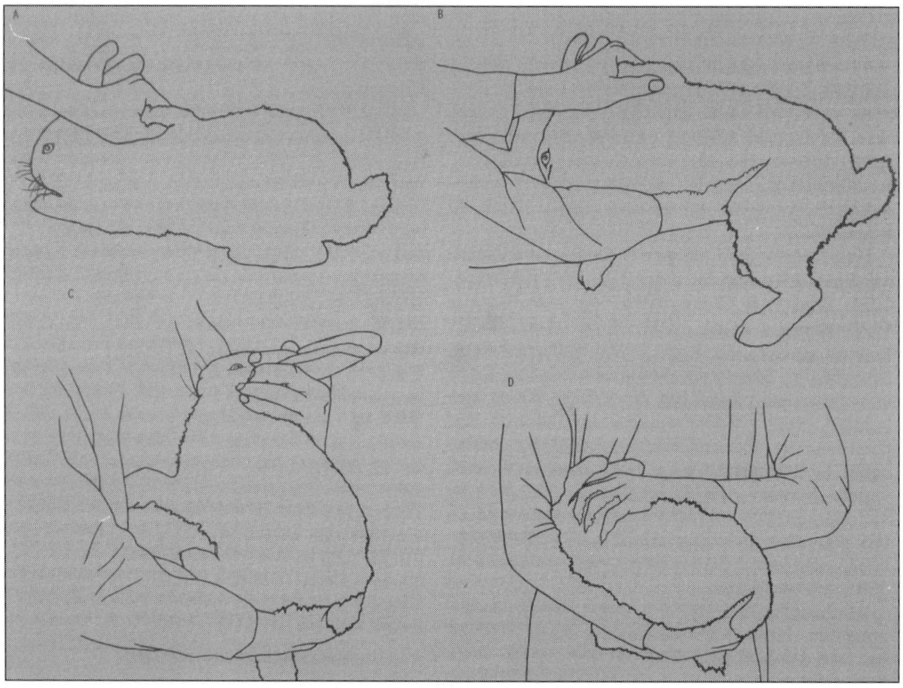

Fig 3.1.5: Handling of rabbit.

3. *Hamsters*
   Golden hamsters often resent handling and may turn over on their back and attempt to bite the handler. Aggressive behaviour is more likely in animals, which have not been handled frequently. Restraint can be achieved by gently, but firmly pressing down on the animal's back with the palm of the hand. Then, the animal is grasped by the scruff of its neck between the thumb and forefingers (Fig 3.1.6). Additional skin along the back can be included in the grasp to perform various manipulations.

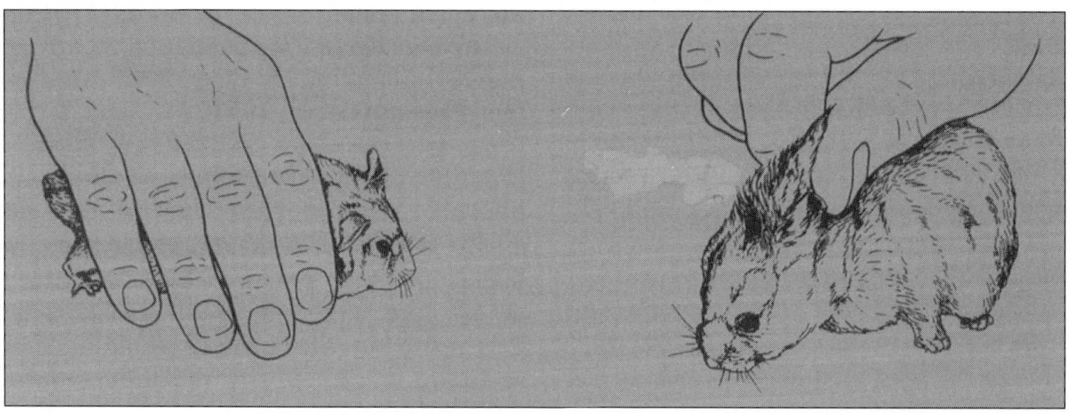

Fig 3.1.6: Handling of hamster.

4. *Guinea Pigs*
   Guinea pigs are easily startled, but rarely attempt to bite if handled properly. They can be lifted from the cage by grasping the thorax with one hand and cupping the other hand under the animal's rear quarters. The rear quarter support is especially

critical with adult animals and pregnant females. If the index finger of the hand grasping the thorax is placed in front of the front legs the animals is less able to jump out of the grasp. Care must be taken to avoid squeezing the chest since this may restrict breathing and cause the animal to panic. It is also possible to fracture the guinea pig's liver by grasping too firmly around the caudal thorax or cranial abdomen while the animal struggles. The "gun-in-holster" method described for the rat can be adapted for the guinea pig (note the absence of a tail to grasp).

5. *Nonhuman Primates*
Chemical restraint of primates is recommended for personnel safety, to help prevent injuries or exposure to zoonotic diseases. This usually involves restraint inside the "squeeze-cage" and IM injection of ketamine for sedation or anaesthesia. All primate handlers must undergo special training and wear the necessary personal protective equipment (PPE) before being allowed to handle primates. Key components of necessary PPE include, but are not limited to gloves, a sleeved gown and face mask.

   With special training unsedated primates weighing two kilograms or less can be caught and restrained by hand (protected with leather gloves). This involves an assistant holding the primate's arms behind its back with one hand and holding and extending the rear legs in the other hand. Commercially available pole-and-collar and chairing devices can also be used for primate restraint.

6. *Pigs*
Methods for restraining pigs include the use of hand restraint, hog snares, pig boards, or slings. With all methods, one can expect a great deal of loud vocalization. Hand restraint usually involves grasping and lifting both rear legs while trapping the neck and head between an assistant's knees (Figs 3.1.7 and 3.1.8).

   Hog snares, which are made of cable or rope with or without a pole are commercially available or can be homemade. A loop of rope/cable is placed inside the pig's mouth and tightened around the maxilla (snout) behind the canine teeth. Various manipulations can be performed while the pig is preoccupied with pulling back against the snare. Commercially available or home-made pig boards are broad, flat sheets of plywood (painted and sanitisable) or durable plastic with two or more oblong handles cut into the top edge of the board. The pig board is used to push the pig against the wall with its head toward the corner of the pen. The handler's legs maintain the pressure against the pig, leaving at least one hand free for manipulations. Pigs can be restrained for manipulations by suspending them in a commercially available or homemade sling that has holes for each leg. The sling may also have a hole to allow access to the ventrum of the pig's neck.

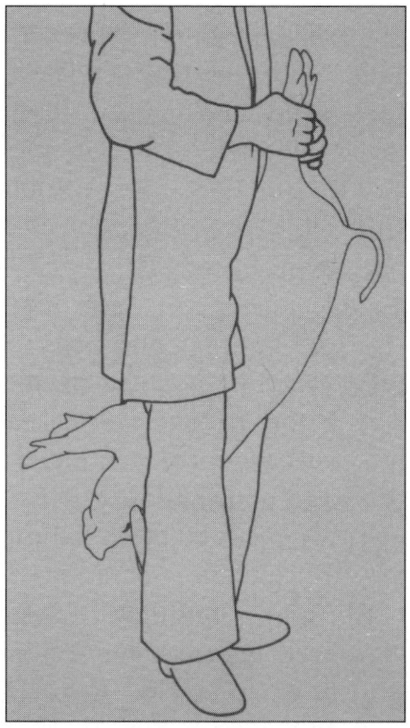

Fig 3.1.7: Hand restraint method for small pigs (side view).

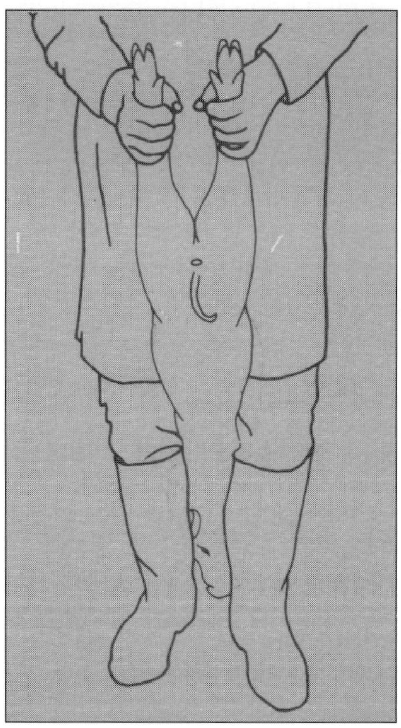

Fig 3.1.8: Hand restraint method for small pigs (front view).

7. *Small Ruminants (Goats and Sheep)*
   If the small ruminant has horns these must be grasped first with one hand to prevent injury to the handlers and to assist with restraint. The animal can be restrained for manipulations by an assistant encircling the head and neck with one arm and backing the animal, rear end first, into a corner with its body pressed against the wall

(Fig 3.1.9). If the other hand is not required to secure the horns, it can be used to help further secure the restraint by placing it under the animal's tail and pressing dorsally and cranially. Fig 3.1.10 shows the method of restraint for sheeps.

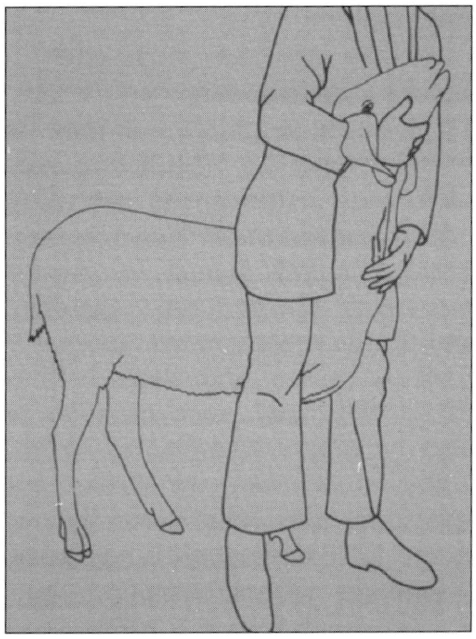

Fig 3.1.9: Handling of goat.

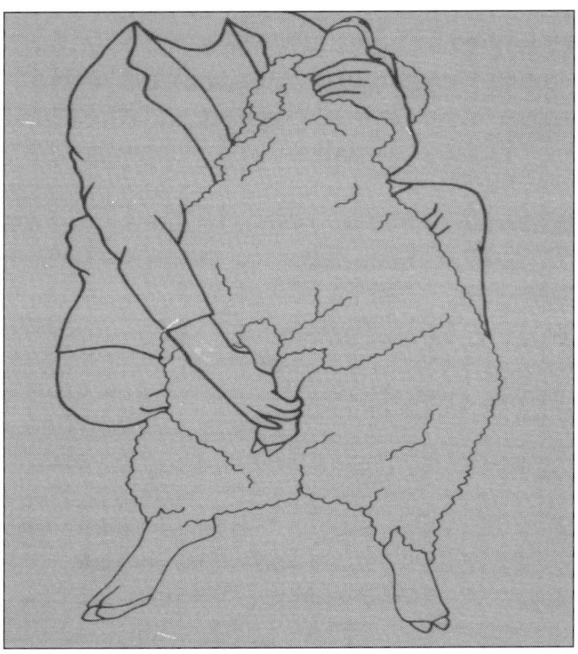

Fig 3.1.10: Handling of sheep.

Oral Dosing/Gavage

Oral dosing with pills or liquids placed in the oral cavity can be unreliable since it is easy for many animals to spit out all or part of the material being administered. Voluntary ingestion and or swallowing is generally dependent on the palatability of the carrier material, such as a favourite food or liquid. Key to successful oral administration of pills is placement at the back of the oral cavity near the base of the tongue where involuntary swallowing reflexes are stimulated. Various pill administration devices are commercially available or can be home-made using appropriately sized plastic syringes with the luer-end cut off and the rough-cut edges heat polished. Gavage is a term used for a method of oral dosing that employs some type of tubular device (rigid or flexible) passed through the oral cavity and down the oesophagus to dose the animal with liquids into the stomach. The diameter of the tube must be smaller than the diameter of the animal's oesophagus. The length to be passed down the oesophagus should not be longer than the pre-measured distance between the tip of the animal's nose and the animal's last rib. Care must be taken to prevent passage into the animal's larynx and trachea, since inadvertent administration of most liquids into the lungs can be fatal. If too much force is used to pass the tube, the oesophagus may be punctured which could lead to the animal's death within days of the event. Species-specific tips for successful gavage are described below.

1. *Mice and Rats*
   Rigid or flexible gavage needles with blunt ends are commercially available for rodents. The type with an enlarged, ball-shaped tip is considered safer since the diameter of the tip is generally too large to be accidentally passed into the laryngeal opening, but still small enough to fit down the oesophagus. The selection of a curved versus a straight or a rigid versus a flexible gavage needle is a matter of personal preference. The gavage needle is attached to the pre-filled syringe in preparation for dosing. The animal must be securely restrained with the head exposed and pointed toward the ceiling, allowing access to the animal's mouth. The ball-tip of the gavage needle is placed behind the animal's incisors and directed along the roof of the mouth towards the back of the oral cavity. The gavage needle can be used to put gentle dorsal pressure on the head to extend the neck and provide a straighter path down the oesophagus. This will help in efforts to avoid the epiglottis, as will a slight, momentary movement of the tip of the gavage needle laterally while advancing past the laryngeal region. If the animal struggles excessively, appears distressed, or if a "dead end" appears to be encountered the gavage needle should be withdrawn and the process restarted. Once the gavage needle is advanced to the pre-measured distance, the liquid should be administered and the gavage needle gently withdrawn.

2. *Rabbits*
   An 8 to 10 french, red rubber urinary catheter is generally used for gavage dosing of rabbits. A speculum to protect the catheter from the rabbit's teeth can be home-made from a syringe case. A small container (cup or beaker) of water should be at hand to verify safe placement of the catheter. The rabbit is restrained in a towel or cat bag and the speculum is placed in the mouth prior to passing the catheter. The catheter is passed through the speculum and advanced down the oesophagus to the pre-measured length. The free end of the catheter is placed in the container of water to verify that the placement is not in the trachea. If air bubbles appear in the water from the catheter corresponding to the rabbit exhaling or following brief manual compression of the chest, the catheter should be withdrawn and the passage attempt repeated. Further verification of correct placement can be accomplished by using a

stethoscope to listen for gurgling sounds in the stomach as air is pushed down the catheter with a syringe. Once correct placement is assured, the pre-filled syringe is attached to the catheter and the liquid instilled. The catheter may be flushed with water to ensure complete dosing. The speculum should not be removed from the mouth until the catheter has been withdrawn. To prevent backflow of liquid, either the syringe should be left on the catheter or the catheter should be pinched during withdrawal.

3. *Hamsters*
   Gavage dosing of hamsters is done using the same method as for mice and rats. If the passage attempt strays too far laterally and the cheek pouches are entered inadvertently, the gavage needle should be withdrawn and the attempt repeated.

4. *Guinea Pigs*
   It is generally not advisable to attempt gavage of guinea pigs due to the presence of the palatal osteum, which makes passage of a gavage needle difficult. Special training would be necessary to make this a feasible methodology.

5. *Nonhuman Primates*
   Commercially available paediatric feeding tubes can be used for gavage in primates. In addition to restraint of the limbs, the head must be restrained from behind with a gloved hand. The security of the head restraint can be improved if the handler can include the zygomatic arch in the grasp. The mouth can sometimes be opened by pushing the buccal wall against the molars, but a bar-type speculum placed in the mouth can be more reliable and safer for the handler. The method and principles of passing and verifying correct placement is similar to that described for the rabbit. Care must be taken to keep hands a safe distance away from the animal's mouth.

6. *Pigs and Small Ruminants (Goats and Sheep)*
   Flexible tubing passed through a tubular metal speculum should be used for gavage in pigs and small ruminants. Restraint for pigs for gavage involves the assistant lifting the front legs and straddling the animals. Restraint of small ruminants is as previously described. The speculum is generally held in place by the person holding the head, which may in some cases be the same person passing the tube. Lubrication of the tube with KY jelly helps ease passage. Correct placement can be verified as described for rabbits.

Injections

The ability to administer materials by injection is essential for most experimental studies employing laboratory animals. Anesthetics and test compounds must frequently be administered to animal subjects by injection. Consideration should be given to attempting disinfection of the injection site by wiping with alcohol prior to penetrating the skin with the needle. There are five commonly used routes of parenteral administration: subcutaneous (SC), intraperitoneal (IP), intravenous (IV), intradermal (ID), and intramuscular (IM). Not all techniques are appropriate for each species. For example, IM injections are avoided in the mouse because the amount of material that can be injected into the mouse's limited muscle mass is so small that the technique is not practical. IP injections are almost never administered to rabbits, as other techniques are more suitable.

It is essential that the appropriate parenteral site is selected. Systemic absorption and distribution differ considerably between sites. Dosage and volume of material administered must be carefully considered relative to the type of agent, site of injection and species used. The size of syringe and needle must also be considered. In order to assure the delivery of an accurate volume of injected material, the volume of the syringe should, in general, not exceed the volume of material to be administered by 10 fold. The length of the selected needle should be long enough that sufficient tissue penetration is achieved but not be so long that it becomes unmanageable or is likely to be inserted to far. The needle's size should be as small (highest gauge) as possible to limit tissue trauma but be large enough so that the injection can be made relatively rapidly and without applying excessive pressure to the syringe plunger. Syringe and needles should generally be of the locking type in order to prevent accidental dislodgement, which may result in autoinoculation or back spray. Proper disposal of used needles and syringes is essential. Needles should never be recapped, as the risk of accidental injection is highest during recapping, and they should always be disposed of into a designated sharps container.

Injection volumes provided in this document are general recommendations. Under some circumstances it may be inappropriate to inject the recommended volume. For example, volumes should be reduced when the agent is irritating or hypertonic. Volumes may be increased when giving isotonic fluids for rehydration and fluid maintenance.

The practice of aspiration, pulling back on the plunger when the needle has been advanced into the injection site, should be employed whenever possible prior to injecting. The decision of whether to continue with the injection, to reposition the needle prior to injecting, remove the needle without injecting, or consider the solution contaminated depends on what is aspirated into the syringe. The handler should develop the ability to control the syringe with the same hand for aspiration and injection.

Intradermal Injection: General principles of intradermal injections include shaving the fur over the intended intradermal injection site, limiting the injection volume to 0.05 ml per site (up to 0.1 ml for larger animals), using a 25 to 27 gauge needle, stretching the skin at the site, inserting the needle nearly perpendicular to the skin with the bevel up, and observing a bleb when the injection is given. Intradermal injections are generally administered on the dorsum in most species, though ID tuberculosis testing in primates may be done in the eyelid or in some cases on the ventral abdomen.

1. *Mice*

   Subcutaneous injection

   SC injections can be administered easily to mice. The needle is inserted between the folds of skin into the base of the triangle that is formed when traction is applied to the animal's scruff. The syringe's plunger should be retracted to verify that a vacuum is created and no blood or tissue fluid can be aspirated. Subsequently, the plunger is depressed releasing the material. In general no greater than 1 ml should be injected per SC injection site in adult mice (> 25 grams). Several sites over the animal's back should be used if larger volumes must be administered. In general, needles should be 0.5 to 1 inch long and 23 or larger gauge.

Intraperitoneal injection

The administration of material into the peritoneal cavity is frequently performed in mice. The aim of this technique is to administer material into the space surrounding the abdominal organs, avoiding injection directly into an organ. Mice should be restrained and held with their ventrum exposed and head pointed downward, this causes the freely moveable abdominal organs to move towards the animal's diaphragm making accidental puncture of organs less likely. A 1 inch 23 or larger gauge needle is inserted into the abdominal cavity in the lower right or left quadrant to avoid the cecum and urinary bladder. The needle should be directed towards the animal's head at an angle of 15 to 20 degrees and inserted approximately 5 mm. Aspiration should be attempted to ensure that an abdominal viscus (hollow organ such as the bladder or colon) has not been penetrated. If material is aspirated, the syringe should be removed and disposed. Never inject gastrointestinal tract contents or urine into the peritoneal cavity, as a bacterial or chemical peritonitis will likely result. In general the volume of material administered into an adult mouse should not exceed 1 to 2 ml.

Intravenous injection

The veins on the lateral aspect of the mouse's tail are excellent sites for IV administration. The principal function of these veins is for thermoregulation. They will dilate when the mouse's body temperature rises in order to disseminate heat. Application of heat to the whole animal or locally to the tail can be used to cause venodilation making vascular access easier. The mouse should be restrained so that its tail is accessible. A 0.5 inch 25 or larger gauge needle is used. The vein is located, the needle inserted by directing the needle into the vein with its bevel pointing upward at an angle of approximately 20 degrees. The needle is inserted slowly visualizing the needle as it enters the vein. Once the vein's wall has been penetrated the needle should be directed cranially approximately 2 mm. Blood should be aspirated into the needle's hub before making an injection.

During material administration the vein should blanch and no material or swelling should be detectable at the injection site. Material should be administered slowly to avoid vascular overload or rupture of the vein from excess pressure. No greater than 0.5 ml should be administered intravenously to an adult mouse. Pressure should be applied over the injection site by gently holding a cotton pledget or piece of gauze over the injection site for approximately 30 seconds to prevent hematoma formation. Preferably the needle should be inserted into the vein midway down the tail, permitting additional attempts for venipuncture proximally if the initial attempt is unsuccessful.

2. *Rats*

Subcutaneous

SC injections are performed in rats using the same technique as was described for mice with the following differences. The rat is usually restrained on a smooth surface while grasping the scruff. The restraint technique using a cloth over the head and thorax can be employed. The volume of material administered can be increased to approximately 5 ml per site in an adult rat (> 300 grams). Syringe size should be increased proportionately and needles should be 22 or larger gauge.

Intramuscular

IM injections may be performed in the rat. Injection volumes are limited to 0.25 ml site because of limited muscle mass. Either the quadriceps muscles located on the cranial aspect of the femur or the caudal thigh muscles of the femur can be used. Care must be taken to avoid depositing material on or near the ischiatic (sciatic) nerve which runs along the caudal aspect of the femur in the thigh. Therefore the needle should be directed cranially if injecting the quadriceps or caudally when injecting into the caudal thigh. A 0.5 inch, 23 or larger gauge needle should be used. The needle is directed through the skin into the muscle belly approximately 3 to 4 mm. Aspiration should be attempted before injecting to determine that accidental penetration of a blood vessel has not occurred.

Intravenous

IV injection technique for the rat is similar to the mouse. However, the vessels are more difficult to visualize, especially in adult rats. The skin overlying the vessels in adults becomes quite thick, making vascular access much more difficult. For this reason the preferred site for vascular access is near either the distal third of the tail or near the tail base. Injection volumes administered to an adult rat should not exceed 2 ml and large volumes should be administered slowly to avoid vascular overload. The technique describing IV administration and needle size in mice should be followed.

Intraperitoneal

The technique for IP injections in rats is virtually identical to mice. Rats should be restrained with their abdomen exposed and their head held downward. The injection site, method and needle size is as described for mice. Because of their larger size < 5.0 ml of material can be administered to an adult rat.

3. *Other Species*

   Most of the principles/techniques for injections in rats can be adapted for other species. Only species-specific techniques or tips are described below. Recommended volumes and sites for injection are listed in Tables 3.2.2. and 3.2.3.

   - **Rabbits**

     Subcutaneous

     Rabbits have ample amounts of loose skin for SC injections and can accommodate higher volumes of material in the subcutaneous space compared to some other species their size. Larger needles (18 to 20 gauge) can also be used for administering fluids.

     Intramuscular

     In addition to the quadriceps and caudal thigh muscle sites, the lumbar muscles can be used for IM injections in rabbits. The principle of keeping the back flexed during restraint must be followed for this injection.

Intravenous
The marginal ear veins of rabbits are readily accessible for IV injections. Restraint must be secure, since the rabbit will attempt to shake its head as the needle penetrates the skin.

Intraperitoneal
While IP injections are not generally recommended for rabbits, the vertical restraint technique with the head pointed toward the floor was described previously in this chapter. The IP injection site is the lower right quadrant of the abdominal cavity to avoid the cecum and urinary bladder.

- **Hamsters**

Subcutaneous
Hamsters have more loose skin than rats, but the same guidelines should be applied for SC injections.

Intramuscular
IM injections are performed as for the rat, but the restraint of the hamster involves grasping the loose skin as described in the restraint section of this chapter.

Intravenous
The hamster's very short tail cannot be used for IV injections. The saphenous vein on the lateral side of the rear leg may be used after shaving the site.

Intraperitoneal
IP injections (Fig 3.1.11) are performed using the same method as for the rat, but the restraint of the hamster involves grasping the loose skin as described in the restraint section of this chapter (Fig 3.1.6).

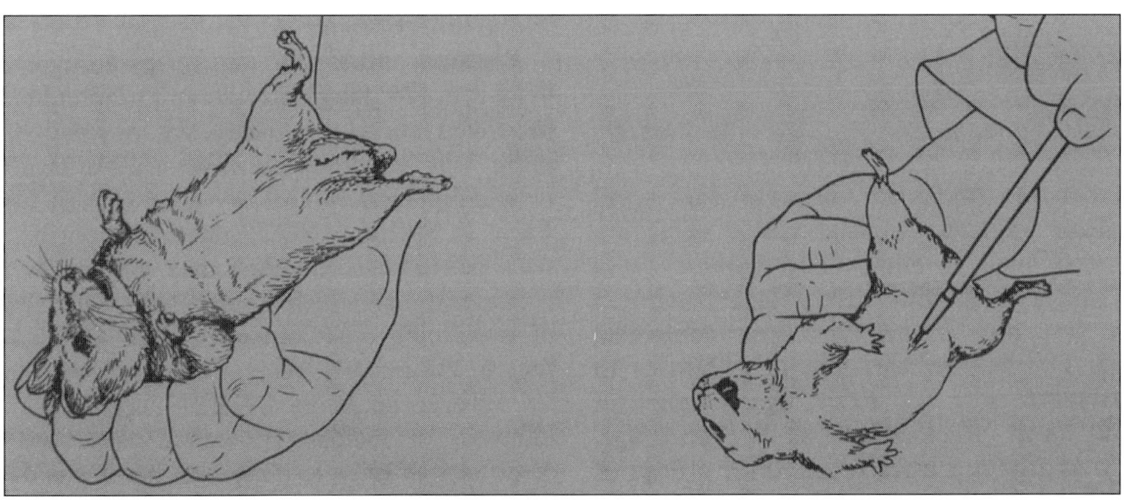

Fig 3.1.11: Method of intraperitoneal injection given to hamster.

- **Guinea Pigs**
  Subcutaneous
  Guinea pigs do not have as much loose skin as most other rodents, particularly over the neck. The skin over the shoulders or thorax can be grasped and lifted for SC injections while the guinea pig is held on a smooth surface. Keep in mind that the guinea pig's natural response is to push up repeatedly when even light pressure is applied to the top of the head.

  Intramuscular
  The techniques for IM injections for rats apply generally to the guinea pig.

  Intravenous
  The guinea pig does not have a tail for IV injections. The saphenous vein on the lateral side of the rear leg may be used after shaving the site.

  Intraperitoneal
  The techniques for IP injections for rats apply generally to the guinea pig.

- **Nonhuman Primates**
  Subcutaneous
  The skin on the torso of the primate may be grasped for SC injections through the cage bars/mesh while the animal is securely restrained by the cage squeeze-back. Handlers should carefully avoid proximity to the animal's mouth or hands.

  Intramuscular
  The cage squeeze-back may be used to restrain primates for IM injections, applying the same precautions for personnel safety as for SC injections.

  Intravenous
  Primate IV injections are given in either the cephalic vein on the anterior surface of the arm or the saphenous vein on the posterior surface of the lower leg. This may be accomplished either in the sedated animal or in the conscious animal while restraining the primate in a specially design cage and pulling the limb through the cage bars. Restraint chairs can also be used to restrain conscious primates for IV injections. Compression of the vein proximal to the injection site will help distend the vein for visualization.

  Intraperitoneal
  Primates are not generally given IP injections, but they should be sedated if IP injections are necessary.

- **Pigs**
  Subcutaneous
  Pigs have almost no loose skin to lift for SC injections except in the flank region. Large volumes of SC fluids cannot be accommodated.

Intramuscular
A large mass of either quadriceps or of caudal thigh muscle is available for IM injections in pigs. An 18 to 20 gauge butterfly catheter is particularly useful, since the flexible tubing helps compensate for the jerking movements of the pig during IM injections. If a regular needle is used there is a risk of breaking the needle off of the hub when the pig jerks. The lumbar muscles can also be used for IM injections.

Intravenous
The pig's ear veins are well-suited IV injections, but the pig should be sedated or if conscious the head must be well secured. A sling can be useful for this procedure. The cephalic vein on the anterior surface of the front leg can be used, but it is difficult to visualize.

Intraperitoneal
It is not generally advisable to perform IP injections in pigs, but if necessary the guidelines used for rabbits can be applied.

- **Small Ruminants (Goats and Sheep)**
  Subcutaneous
  The loose skin on the side of the neck, above the shoulders, or at the flank can be lifted for SC injections.

  Intramuscular
  Either the quadriceps or caudal thigh muscles can be used for IM injections.

  Intravenous
  Small ruminants have large bilateral jugular veins along the antero-lateral surface of the neck in a groove lateral to the trachea. Compression of the proximal portion of the vein at the base of the neck will distend the vein for visualization. Sheep will usually need to be shaved over the site for visualization (Fig 3.1.12).

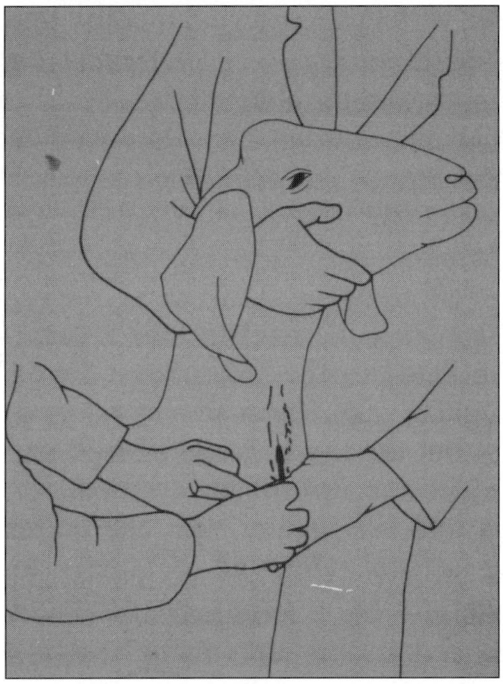

Fig 3.1.12: Intravenous injection given through jugular vein of goat.

Intraperitoneal

IP injections are not recommended for small ruminants.

CHAPTER **3.2**

BLOOD COLLECTION FROM LABORATORY ANIMALS

Jason Villano

Blood collection, often referred to as "bleeding" the animal, is one of the routine procedures done in a laboratory animal research setting. Thus, proper blood collection plays a significant role in animal welfare. The handling and bleeding process can induce stress and possibly pain to the animals. Hence, every effort must be taken to reduce this stress as it may alter and affect their physiology and behaviour leading to variations in the research results. Acclimatization and adaptation of the animals to their new surroundings and gradual introduction to certain procedures such as blood collection will greatly assist the animals in settling down.

The following is a list of guidelines for safe blood withdrawal in laboratory animals:

1. Variations in animal species, age, and gender reflect different blood volumes in milliliters of blood to kilogrammes of body weight. For most species, this blood volume is equal to approximately 6 to 8 % of the body weight. Blood volume removed is usually replaced within 24 hours, but replacement of blood cells takes longer. For normal healthy adult animals, it is usually safe to assume that the red blood cell renewal occurs on a 14 to 21-day cycle and other blood cell constituents normally take up to two weeks to recover.
2. All personnel should be careful in handling animal that is aged, stressed, or has undergone experimental manipulations, as these factors may be deleterious to the animal's health and well-being.
3. Most animals will go into shock if 25 to 30 % of their blood volume (or approximately 2 % of the body weight) is collected over a short period of time. In many animals, removal of 30 to 40 % of the blood volume may cause death. For

these reasons, a general thumb-rule is that blood removal should not exceed 10 % of the blood volume in a single bleed and no more than 20 % over a two-week period. Another thumb-rule accepted at many institutions is that for a single blood draw, 0.9 to 1 % of the animal's body weight can generally be removed while 2 % may be allowed if intravenous fluid replacement therapy (warmed isotonic fluids) is administered as the blood is withdrawn. For a 1 to 2 % of body weight blood collection, a two-week recovery period should be allowed between consecutive collections to allow for renewal of blood cells.

4. Approximately 3 to 4 % of the body weight (50 to 75 % of total blood volume) can be obtained through exsanguinations (a terminal procedure). Giving fluids during bleeding to maintain the animal's blood pressure can increase the total volume of blood cells obtained. Anaesthesia is required prior to exsanguinations in order for this to be an approved method of euthanasia.
5. For chronic blood sampling or blood collection frequency of more than once every two weeks, a total of 0.5 % of the animal's body weight can be removed each week with this total volume being spread out over the entire week if necessary.
6. For maintenance, animals usually require 4 ml/kg/hour of fluids. The volume of fluid replacement such as Lactated Ringers Solution should be equal to three times the volume of blood removed and should be given at the rate approximately equal to the rate of blood loss. Fluid replacement therapy is not usually necessary if the animal is healthy and able to eat and drink immediately after the blood collection of recommended volumes.

The recommended blood collection volumes for a number of species are given in the following table (Table 3.2.1).

Table 3.2.1: Blood collection volumes for selected animal species commonly used in laboratory

Species	**Body weight (BW) in grams**	**Adult blood volume (ml) (6% BW)**	**Single sample volume (ml) (0.9% BW)**	**Multiple sample volume (ml) per week (0.5% BW)**	**Exsanguination volume (ml) (4% BW)**
Mouse	20	1.2	0.18	0.1	0.8
Hamster	100	6	0.9	0.5	4
Rat	250	15	2.25	1.25	10
Guinea pig	500	30	4.5	2.5	20
Rabbit	2000	120	18	10	80
Goat	50,000	3000	450	250	2000
Sheep	60,000	3600	540	300	2400
Pig	50,000	3000	450	250	2000
Monkey	2500	150	22.5	12.5	100

Bleeding Techniques

Blood collection may be performed adequately in alert animals of most species using the appropriate restraint. The proper restraining method is necessary to prevent movement that may result in laceration of the blood vessel or other organs causing serious complications. One possible example is that of diabetic animals, which have impaired healing and revascularization. Using the correct restraining methods also protects the handler, addressing occupational health and safety concerns, especially in the case of blood collection in nonhuman primates.

The following table (Table 3.2.2) indicates the usual blood collection sites and the recommended restraining methods for a number of animal species. Individual animal behaviour and species variation play a very crucial factor in determining the appropriate method of restraint.

Table 3.2.2: Appropriate blood collection sites and recommended restraining methods for selected animal species commonly used in laboratory

Species	Blood collection sites	Recommended restraining methods
Mouse	Saphenous vein (medial or lateral)	Manual restraint
	Mandibular (facial artery/vein)	Manual restraint
	Retro-orbital sinus/plexus sampling	Anaesthesia
	Cardiac puncture*	Anaesthesia
	Tail vein, tail arteries, tail clip (Fig 3.2.3)	Restraint device and/or manual restraint
Rat	Saphenous vein (medial or lateral)	Restraint device and/or manual restraint
	Jugular vein and cranial vena cava	Manual restraint or with anaesthesia
	Retro-orbital sinus/plexus sampling	Anaesthesia
	Cardiac puncture*	Anaesthesia
	Tail vein, tail arteries, tail clip	Restraint device and/or manual restraint
Guinea pig	Crania vena cava	Anaesthesia
	Ear vein	Restraint device and/or manual restraint
Rabbit	Ear artery and vein (Fig 3.2.1)	Manual restraint
	Cardiac puncture*	Anaesthesia
Goat and sheep	Jugular, cephalic, saphenous veins and other superficial vessels	Manual restraint
Pig	Ear veins, vena cava, jugular vein (Fig 3.2.2)	Manual restraint (small animals) Chemical (big animals)
Monkey	Femoral, saphenous (Fig 3.2.4), and cephalic veins	Chemical (>2 kg BW) Manual restraint (<2 kg BW)
Dogs and cats	Jugular, cephalic, femoral, saphenous veins and other superficial vessels	Manual restraint

Other guidelines to be observed include:

1. The site of blood collection should be disinfected or at least swabbed with ethyl alcohol as the skin, the body's primary defense, will be breached.
2. *Cardiac puncture may only be used for terminal blood collection due to possible cardiac tamponade, pulmonary hemorrhage, and pneumothorax, which can be potentially fatal to the animals.
3. Tail snipping in neonatal and juvenile mice should be followed by sealing of the tail with wax, or other measures, to halt bleeding and prevent maternal cannibalism.
4. Blood withdrawal from superficial vessels such as those in the ear and tail may be facilitated through dilation of the vessels by:

   - increasing body temperature with the use of a heat lamp, an increase of room temperature, or application of hot damped cloth or gauze.

- using tranquilizer or local anaesthetic cream.
- gentle finger-tapping of the skin overlying the vessel.

5. An animal must not be returned to its cage until complete haemostasis has been achieved using gauze and direct digital pressure. Arterial punctures may require up to several minutes of pressure.
6. It is possible to evaluate whether an animal has sufficiently recovered from blood collection by monitoring the haematocrit (or packed cell volume – PCV). However, it is important to note that PCV will not show alterations acutely, but will manifest changes after the body has replaced the volume of blood extracted with fluids.

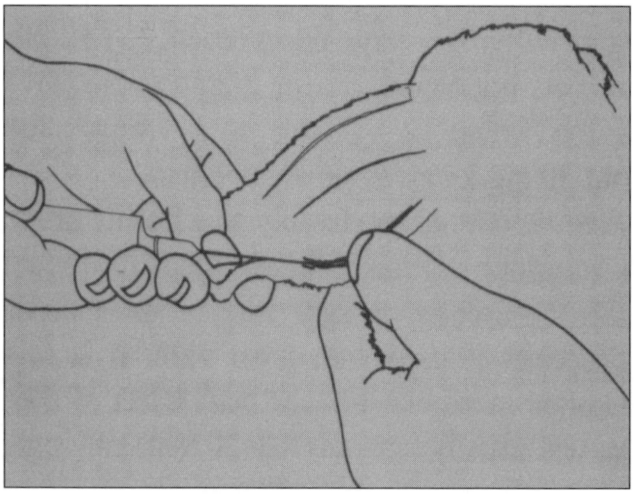

Fig 3.2.1: Blood collection from rabbit ear.

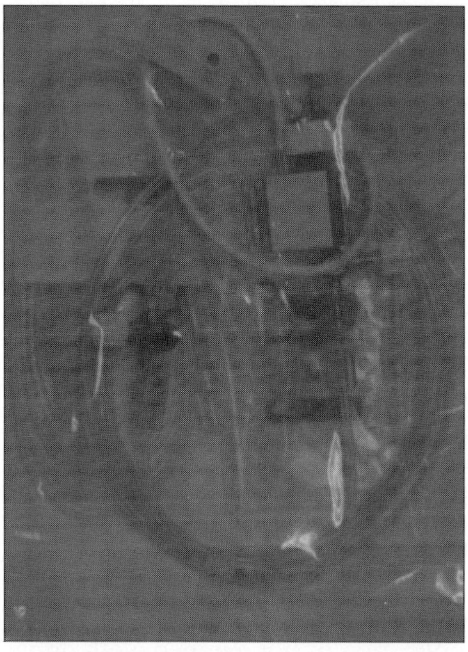

Fig 3.2.2: Pig Jugular Vein Catheterization Kit.

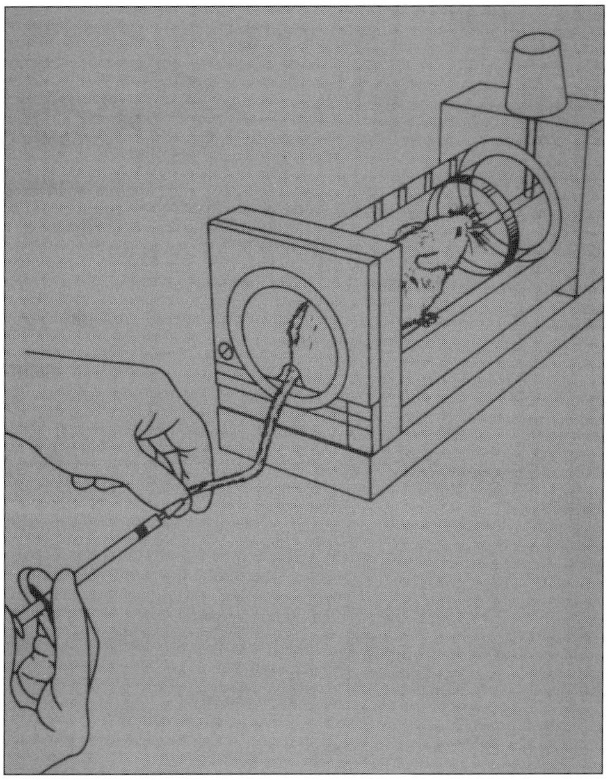

Fig 3.2.3: Blood collection from tail vein of mouse.

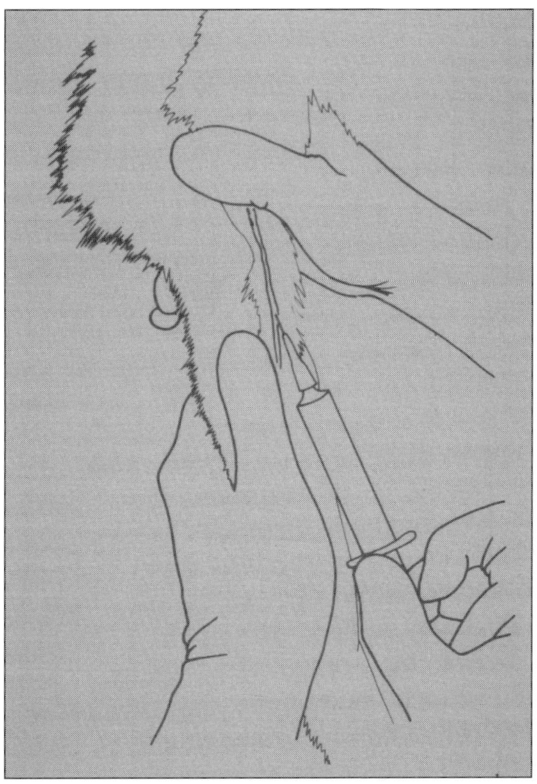

Fig 3.2.4: Blood collection from saphenous vein (monkey).

CHAPTER **3.3**

ANTIBIOTIC COVERAGE AND THERAPY

Darvi Sergio

When an infection is suspected in an experimental animal and antibiotic therapy is to be initiated, the principal investigator, the veterinarian and veterinary staff must be familiar with the organisms that are likely to be encountered during the course of the study. The selection of the agent or agents is based on the history, the physical examination, the host defense status, the site of infection, the overall clinical severity of the infection and the response of the host.

If antibiotic therapy is to be successful it must be started as soon as infection is suspected in an animal or in a number of animals in the experimental group. Antibiotic therapy is initiated with an agent or combination of agents whose action is broad enough to cover all the suspected microbial pathogens. The application of such broad-spectrum antibiotic therapy in the absence of microbiologic confirmation is termed as empirical therapy. Definitive therapy is initiated after the host response to the infection and to the empirical treatment has been monitored and the results from the microbiology laboratory have been assessed.

Factors Influencing Application of Antibiotic Therapy

A. *Host Factors*

Host can influence the application of antibiotic therapy. *Hypersensitivity to antibiotics* is one of them, for example, penicillins are fundamentally toxic to guinea pigs. *Concurrent illnesses* in animals can also influence antibiotic therapy. For example, animals with immunosuppressive illnesses are vulnerable to opportunistic pathogens. Animals, which have developed renal insufficiency and/or diabetic models for research, may be unusually susceptible to direct drug toxicity.

Teratogenecity is a major concern when any drug is administered during pregnancy. For this reason, the use of chloramphenicol and tetracyclines should be avoided in pregnant animals.

B. *Routes of Administration*

Some antibiotics are absorbed sufficiently via oral route while others are impeded by food and medications (example, antacids). There are also antibiotics that cannot be given intramuscularly because of local pain or necrosis at the injection site. Some small animals should not receive intramuscular injections due to their small muscle mass and the potential for sciatic nerve damage.

There are various routes that can be employed for drug delivery to the animal. They include:

1. Enteral Therapy — the desired effect is systemic and given via the digestive tract. This therapy includes the following methods:

   - **Administration via water**
     This may be the easiest and least expensive method for treating groups of animals; many animals can be treated simultaneously with minimal stress. However, the disadvantage is uncertainty of the exact dosage received by individual animal due to variables like disease status, age, sex, body weight, dehydration, which may influence water intake. Antibiotic preparations should be water soluble and stable, palatable and well absorbed from the gastrointestinal tract.

   - **Administration via food**
     Some drugs can be easily hidden in a piece of feed and the animal may eat it voluntarily with no stress of handling. Customized feed may also be available with antibiotic already mixed in.

   - **Direct oral administration**
     This may be the preferred method because the amount administered is placed directly into the animal's mouth, preferably to the back of the tongue. With oral administration, palatability is helpful, but alternatively a gavage tube may be used to instill the drug directly into the stomach. This would assure that the intended dose is ingested. Proper restraint and dosing techniques are necessary to prevent regurgitation, gagging out or aspiration.

2. Parenteral Therapy — indicated for antibiotics that are not very well absorbed from the gastrointestinal tract, or if rapid effect is required. Most commercial drugs are too concentrated, and accordingly needs to be diluted before use especially on small animals. Parenteral methods include:

   - **Subcutaneous administration**
     This method minimizes tissue damage to muscle and large volume of drugs can be injected.

- **Intramuscular route**
  This method should be rarely used in small animals, such as mice, due to their size, small muscle mass and potential for sciatic nerve damage.

- **Intraperitoneal route**
  Easy administration compared to other parenteral routes. However, person carrying out the injection should be cautious not to hit any vital organs in the peritoneal cavity.

- **Intravenous route and intraosseous routes**
  Are the most challenging and generally reserved for severe cases of bacterial disease. These routes are occasionally used for drugs and fluids in emergency medicine.

3. Topical Administration
   Topical administration has local effect. Antibiotic is applied directly where its action is desired. However, it has limited use in rodents due to their fastidious grooming habits.

Use of Antibiotics in Surgery

Antibiotics are poor substitutes for good surgical technique. Proper hand scrubbing and gloving by the surgeon, preparation of the operative site (clipping hair and disinfection), sterile techniques, proper wound closure and minimizing tissue trauma and operating time are key considerations in preventing infection. If antibiotics are indicated, a single dose should be given parenterally approximately two hours prior to the first incision or intravenously preoperatively allowing sufficient time to obtain therapeutic blood and tissue levels. Giving the first dose by any route in the middle or at the conclusion of a surgical procedure is not as effective. Antibiotics are indicated with the following risk factors:

1. Procedures of more than two hours in length.
2. Abdominal (bowel) surgery.
3. Severe blood loss.
4. Diabetes mellitus.
5. Foreign body implantation.
6. Cardiac valvular disease or valve surgery.
7. Debilitation.
8. Trauma/infection cases.
9. Immunosuppression.
10. Gross contamination during surgery.

Treatment beyond the perioperative period may not be necessary and it has been shown to be no more effective than a single dose given two hours preoperatively.

Guidelines for Prudent Antibiotic Use

Often it is the veterinarian and the animal caregiver that decide how antibiotics will be used, but the principal investigator should be consulted. Everyone who administers antibiotics to animals should understand and be willing to apply these general concepts of prudent use. Below is a list of guidelines for prudent antibiotic use in animals.

A. *Provide a system of care to prevent common disease*
It is more cost effective to prevent disease than rely on antibiotics to treat disease once it has developed. Minimizing disease risk is a necessity for any research animal facility and is achieved by maintaining good sanitation and hygiene, providing high quality feed, implementing biosecurity measures, performing regular health exams, and using vaccines and parasite control.

B. *When animals do get sick, an accurate diagnosis should be obtained*
This ensures that antibiotics will be used to treat the appropriate clinical indications. Diagnosing an animal's condition should be based on clinical signs, history, necropsy results, laboratory data, and past experience. The veterinarian can help provide an accurate diagnosis.

C. *Determine which class of antibiotics are the most appropriate option*
Keep in mind the treatment outcomes you want for the animal and what types of therapy will help you achieve them. Some infections respond better to antibiotics that are bacteriocidal, rather than bacteriostatic. Some antibiotics penetrate specific tissues better than others. There may be an advantage to combining two classes of antibiotic, but care must be taken to avoid known incompatibilities.

D. *Choose the most appropriate antibiotic for the circumstances*
Not all antibiotics work the same and some infectious organisms develop resistance to certain antibiotics. Culture and sensitivity tests of the causal organism will help determine the type of antibiotic most effective in clearing the infection. Select an antibiotic form that can be easily administered and will result in effective concentrations reaching the site of infection. When appropriate, use local therapy instead of systemic. When possible use only medications approved for the intended use.

E. *Work with your veterinarian to enhance therapeutic options*
Veterinarians have the knowledge and resources necessary to determine the most effective therapy. They can offer valuable guidance for the prudent use of readily-available antibiotics and can offer enhanced therapy through the use of prescription and extra-label use medications. They may provide written protocols for diagnosing and treating common disease conditions.

F. *Use antibiotics and other medications as ordered*
Use medications at the ordered dosage for the appropriate species and indication. Train all personnel involved with antibiotic use and animal care on disease indications, dosages, routes of administration, injection site precautions, treatment

duration, withdrawal times, storage, handling, record keeping, and accurate diagnosis of diseases common for your operation.

G. *Treat the appropriate animals*
Limit therapeutic antibiotic treatment to those animals that are sick or are legitimately at-risk of becoming sick. Avoid prolonged treatment of animals and consider salvage alternatives for chronic cases or those with a poor chance of recovery.

H. *Store antibiotics and other medications appropriately*
Drug integrity is maintained by following label and veterinarian instructions for proper handling, storage, and observation of expiration date. Mistakes in administering the wrong medication are less likely if drugs are clearly labelled and stored in the appropriate places.

I. *Minimize environmental contamination*
Dispose of outdated medications according to label directions or veterinary advice. Use disposal methods that minimize contamination of soil and water supplies. Provide feed and water medication delivery so that there is minimal spillage into the environment.

J. *Use records to track treatments and evaluate outcomes*
Keep accurate, detailed, and current records of antibiotic treatments and outcomes. Identify all animals either by individual, cage, or group so they may be monitored. Treatment records should include the identity of animals treated, dates treated, drugs administered, personnel that administer the drugs, and the amount administered.

CHAPTER 3.4

ANIMAL PREPARATION AND TRANSPORT

Robert Ng

Sourcing and Transportation

The use of physiologically stable and healthy animals contributes to successful research outcomes due to reduced stress and health-related variables. The prerequisite is therefore to source for appropriate animal models from accredited vendors familiar with local guidelines for animal transportation. Although some transport related stress is inevitable, animal transportation should be planned to minimize transit time and risk of zoonoses and to protect against environmental extremes. Whenever possible, the transport vehicles should be designated for animal use only and be disinfected after each delivery.

Rodents are housed in air-conditioned facilities. The transfer vehicle should thus also be air-conditioned and a van with panel cargo carriage space as a secondary barrier will be appropriate. For big animals like pig, sheep and monkeys, a lorry with a fan ventilated cargo carriage space that is enclosed with canvas canopy or stainless steel meshed cover will be acceptable for animal comfort in Singapore's context.

Rodents are normally transported in groups within waterproof paper carton boxes that have filtered side windows with punched holes for ventilation. Rodents supplied from overseas sources need to include, within the containers, pellet food and nectar sachet for drinking water. Large animals are contained individually within locked stainless steel meshed trolley enclosure configured sufficiently for freedom of movement. Large animals, such as macaques, coming from overseas are normally housed singly in wooden crates.

Quarantine and Acclimation

Quarantine is the isolation of animals of unknown health or microbial status to minimize the chances for introduction of pathogens into an established colony. The AVA has established quarantine guidelines for the Research Institution to comply with and implement, especially for animals imported from overseas. Imported experimental animals are normally from AVA-accredited sources and are mostly purpose-bred for research and come with health certificates. The onus is for the Research Institution to determine the need for quarantine. The animals are required to be housed individually in isolated containment for health evaluation and treatment. Where designated containment is unavailable, animals should be separated to a secluded end of the animal room, preferably with dedicated waste discharge to prevent contamination to other animals.

- Nonhuman primates (NHP) require a quarantine period of one month. Intradermal tuberculosis testing will be done at two-week intervals and blood samples will be evaluated for SRV, SIV, STLV and Herpes B viruses.
- Pigs arriving at the dedicated breeding colony of the Department of Experimental Surgery (DES), Singapore General Hospital will be tested to confirm that they are free from Classical Swine Fever, Aujesky Disease, Nipah virus, Brucellosis, Leptospirosis and Toxoplasmosis. Prophylaxis with either gentamycin (2 mg/kg) or oxytetracycline (7 mg/kg) will be given intramuscularly daily for three days against pathogens such as *Streptococcus suis.*
- Sheep will be quarantined for a period of time for evaluation of Q Fever status.
- Rabbits arriving at DES will be given Protexin antibiotics (Lactobacillus, Bifidobacterial, *Streptococcus salivarius* and *Enterococus faecuim*) through oral drinking water at a dose of 3 g/ml for three days. Ivermectin at 300 ug/kg will be given subcutaneously every two weeks up to six weeks to eradicate parasites.
- Dogs are seldom used for research and they are normally acquired from the pounds and will be quarantined for at least two weeks for deworming with Drontal Plus (50 mg praziquantet, 144 mg pyrantel embonate and 150 mg febantel) at a dose of 1 tablet per 10 kg. External parasites are treated with Fipronil spray and animals are vaccinated against distemper, adenovirus, parainfluenza, parvovirus, leptospirosis. Dogs from the pound have high incidence of heartworm disease and should be tested with a heartworm antigen test and possibly treated with Immiticide (melarsomine dihydrochloride) at 2.5 mg/kg (intramuscular) twice 24 hours apart.
- Rodents are mostly purpose-bred for research and have been serologically screened against viral agents like Mouse Hepatitis virus, Hataan, Sendai and Parvovirus.

Regardless of duration of quarantine, newly received animals should be given appropriate acclimation for physiologic, psychologic and nutritional stabilisation before their use. The animals are housed in specie-specific rooms to prevent interspecies disease transmission and to eliminate anxiety and possible behavioural changes. In DES, the minimum period of acclimation for rodents is three days, pigs three to five days and sheep, rabbits and monkeys for one to two weeks prior to use for surgery or non-surgical procedures.

CHAPTER **3.5**

PREPARATION AND IMPLEMENTATION OF ANIMAL SURGERY

Robert Ng

To achieve optimized outcome of surgery, considerations should be given to presurgical preparation, provision of sterile environment, atraumatic surgery, maintenance of animal physiologic status and smooth recovery. Aseptic technique is always advocated for all animal species to reduce microbial contamination to the lowest level. The use of antibiotic should also never be considered as a replacement for aseptic procedure.

Generally surgical procedures can be categorised into four classes: minor survival, major survival, multiple survival and non-survival. Minor survival surgery does not expose a body cavity and causes little or no physical impairment as in wound surturing or subcutaneous cannulation. Major survival surgery penetrates and exposes a body cavity and/or produces substantial impairment of physical and physiologic function as in laparotomy, thoracotomy, craniotomy, amputation and hypophysectomy. Multiple survival surgery may involve the initial creation of a surgically induced animal model with physiologically compromised or diseased status as in myocardial infarction, cirrhotic liver or tumour model to be followed up by another surgery to treat, reverse or improve on the induced impairment. In nonsurvival surgery an animal is euthanized before recovery from anaesthesia as when they are used in surgical skills training courses.

Animal Preparation

1. *Pre Surgery Preparation*
   Pre-surgical preparation of an animal is based, in part, on the specific protocol requirement from the Principal Investigator (PI) and approved by IACUC. Some pre-

surgical procedures are standard with the aim of presenting the animal as healthy and well prepared to withstand the trauma of surgery and ensure its survival. These pre-surgical preparations followed in DES are listed below:

- Overnight fasting with no access to food from evening to the next morning is mandatory in most large animal species to prevent regurgitation. A gavage tube may be inserted during surgery for stomach drainage, especially for ruminant.
- For an animal (40 to 60 kg) that is going for angioplasty, it is recommended that they are premedicated three days prior to surgery with daily oral intake of clopidogrel (75 mg), aspirin (300 mg) and verapamil (240 mg). It is the onus of the PI to determine whether to premedicate a single drug or in combination.
- For bowel preparation as required for colorectal surgery, castor oil (2.5 ml/kg) is given orally via gastrogavage. Care should be exercised to avoid oesophageal reflux especially when given to an unconscious animal, which should have its body inclined in a head up posture. Colyte and biscadyl have also been used for bowel preparation. Alternatively, the animal can be given a colonic lavage preparation of two sachets containing 128 g of macrogol (INN) 400 in one litre of water.
- For ruminants especially goat, 1 g tetracycline can be given subcutaneously a day before surgery for animals weighting 40 to 50 kg to reduce bacterial load in the gastrointestinal tract to avoid abdominal bloating during anaesthesia.
- Animals undergoing orthopaedic procedures usually require X-ray imaging as pre-operative control reference.
- Blood samples are normally taken for complete blood count and clinical chemistry (FBC) as health certification prior to surgery. Blood cross matching is done on animals in preparation for major surgery like organ transplants and cardiothoracic surgeries where blood reserve is sometimes necessary as standby for blood loss replacement.

2. *Animal Surgical Preparation*

The anaesthetized and intubated animal is shaved on the intended operative site and hair and fur are removed with a vacuum evacuator that incorporates HEPA-filter exhaust to trap contaminants from discharging into the room. The shaved area is swabbed with 1 % centrimide followed by 0.05 % chlorhexidine (or other suitable disinfectants such as iodine). The animal is then weighed and rodents are transferred to the microsurgery laboratory and large animals to the operating theatre suite. A hydraulic trolley can be used for easy transfer of the animal to the operating table. The operating table has either an inbuilt heated tabletop or a water-circulating heating pad to minimize development of hypothermia. For certain surgical procedures like orthopaedic and angioplasty the fluoroscopy room is used for C-arm fluoroscopic imaging.

The animal is normally placed in a supine position (dorsal recumbancy) to allow easy access to the face, neck, chest, abdomen, pelvis and anterior surfaces of the limbs. The prone or sternal position (ventral recumbancy) is more suitable for operations on spine and cranium. A lateral position (lateral recumbancy) is desired for some chest operation and certain procedures on the liver, kidney, ovary and spine.

An intravenous (IV) infusion line is set up, if required, via cannulation of the ear vein or other vein on the neck (jugular) upper limb (cephalic) or lower limb (saphenous). The fluid bag (parenteral, colloid or blood) is sometimes required to be inserted into a pressure bag for assisted pressurized flow when the cannulated vein is small and exerts flow impedence. Arterial lines may be placed for invasive blood pressure (IBP) monitoring. This is performed under sterile conditions and involves doing a cut-down to access and cannulate the carotid, tibial or femoral artery. For cardiothoracic surgery, a syringe pump may be required for infusion of an intravenous anaesthetic, such as propofol (0.5 ml/kg) to provide replacement anaesthetic when the ventilator is shut down during vascular bypass through the extracorporal heart lung machine. A defibrillator should also be available for emergency intervention. For laparoscopic surgery, a CO_2 insufflator is available for intraperitoneal insufflation at a pressure of 10 to 15 mmHg and for eye surgery the pupil is dilated with tropicamide.

A blood sample is routinely taken for complete blood count (CBC) for baseline health reference. A peri-surgical wide spectrum antibiotic coverage may be given intramuscularly or intravenously. Alternatively, antibiotic may have been given two hours before surgery. The shaved operative areas is then given a final scrub with 1 % povidone iodine. Swabbing should be done in a rotating fashion starting from the centre spot and extending to the perimeter to avoid accidental introduction of contaminations from the hair and fur on the unshaved parts. The animal is draped with sterile surgical towel held in place with towel clips. Sterile instruments are displayed on sterile draped mayo stand and sterile handgrip is inserted onto the operating light for adjustment of light focus. The animal is ready for surgery.

3. *Sterile Scrubbing Up and Gowning*
In the changing room, surgical team will change from street wear into blue scrubs (the scrub colour is not proscriptive) and put on a nonfogging face mask, head cover and shoe covers. An N95 face mask and shield is recommended for use involving bone drilling, burring or cutting and for surgery on nonhuman primates.

In the scrub room, wet the hand and forearm and apply a chlorhexidine based antibacterial cleanser for hand rubbing on the hand and forearm for one minute. Rinse off with water and tear open the scrub brush pack to retrieve a scrub brush that has a sponge impregnated with 7.5 % povidone iodine soap and has a brush on the other. Spread the skin soap evenly with the sponge applicator. Sponge both hands and forearms thoroughly paying particular attention to the nails, cuticles and interdigital spaces. Then use the brush to scrub the nails, fingers and palms. The elbow is used to turn the tap lever for washing with running water. Raise the hands above the elbows so water runs down from the hand to the elbow to prevent dirty water draining into the clean area. This is repeated two or more times for a thorough scrubbing. Hands and arms are dried with a sterile towel. Sterile gowning and gloving follows.

4. *Animal Surgery*
Before undertaking any operative procedure it is essential to have a thorough anatomical knowledge and good surgical technique. Dissection should be as atraumatic as possible, following natural tissue planes and avoiding unnecessary and

excessive handling of tissues. In surgery, the orientation terms used should provide useful reference to operative approach such as cranial (towards the head); caudal (towards the tail); dorsal (towards the animal's back); ventral (towards the animal's front); medial (towards the midline); lateral (away from the midline); contralateral (the other side); proximal (near the body); distal (away from the body); prone/sternal (lying on the face); supine/dorsal (lying on the back) and decubitus (lying on side). Surgical procedures can be described in six steps.

i) Incision

A skin incision can be made with the dominant hand holding the scalpel holder with the thumb opposing the two adjacent fingers as in a pen grip or alternatively, as in a knife grip for cutting food on a plate. The incision made with a scalpel blade must be a single smooth cut to penetrate the subcutaneous tissue and the fascia. Hand pressure required during the cutting manoeuvre depends on the toughness and depth of the skin. Avoid multiple scratch-like movements during the incision. The other hand is used to steady the skin with the index and thumb stretching the skin laterally to facilitate the incision. Bleed points on the skin tissue can be picked up with a forceps and cauterized with diathermy. Muscle may be separated or split using a blunt technique with closed scissors or haemostat to push through the muscle and opened to separate the longitudinal fibres. It is better not to cut across muscle fibres since they are soft and friable and not easy to resuture.

- For abdominal surgery, the organs are conveniently accessed by a midline incision. The subcostal incision running parallel to the last rib below the costal margin on one side is occasionally used to approach the liver or spleen.
- The thoracic cavity can be widely exposed with a ventral midline incision of the skin and the sternum divided with an oscillating saw. Operating on a single lung, the oesophagus or the lymphatic system, the dorsolateral approach is normally used. Incision of the intercostal space of the 4th, 5th or 6th rib will provide good access.
- Most surgical procedures on the neck can be undertaken through a midline incision with the animal lying supine and the head hyperextended. The incision will involve cutting through the skin, underlying fascia and in some cases, the strap muscles.
- For craniotomy, the incision is usually in the midline from above the eyebrows to the vertex and then extended to the appropriate side at one or both ends. The periosteum of the skull is scraped away from the proposed line of bone incision with a periosteal elevator and craniotomy can be performed with a dental burr, trephine and/or a gigli saw.
- In laminectomy, a dorsal midline incision is made to extend one vertebra length below and above the required vertebral site. The lamina is removed with a bone nibbler (rongeur) for exposure of the spinal cord. Any bleeding encountered can be reduced with swab packing.

ii) Retraction

A retractor may be needed to open up the incision by holding back the skin and muscle covering to visualize the body cavity. Various retractors are available which include the Balfour abdominal retractor, self-retaining retractor for small wound visualization and a Finoccetto rib spreader for thoracic cavity exposure. Local or deep recessed tissue retraction can be achieved with a Deaver or cat paw retractor.

iii) Haemostasis and Vascular Occlusion

The control of haemorrhage during an operation is essential not only to prevent blood volume depletion but also to present a clean operative field to facilitate dissection. Bleeding when it occurs may be arterial, venous or capillary. A capillary oozing and bleeding from other small vessels will eventually stop from local clot formation with swab pressure on the bleeding area or with diathermy application. For more extensive bleeding or bleeding into a restricted area, good visualization may only be possible with the use of suction apparatus. If the vessel is large, an artery clamp can be applied and ligature placed with suture tie is necessary. Tying a knot around the tips of a haemostat requires skills of knot tying and in the removal of the haemostat to ensure that the suture does not slip off. Normally three "throws" (half hitches) are required for a secured knot, depending on the type of suture. Alternatively fragile bleeding points or bleeding in an inaccessible area can be controlled with metallic or plastic stapling clips which are commonly used in endoscopic surgeries. For generalized bleeding, which is difficult to control as in organ transection, the application of haemostatic sponge or gel containing absorbable gelatin, calcium arginate or fibrinogen base can be used. Diathermy set to coagulation mode can be used for tissue dissection or one can also use a harmonic knife that uses radiofrequency technology. Vascular occlusion can be achieved using a snare procedure by passing suture around the vessel and suture ends going through a soft plastic tube which when pulled tight together with the tube and when clamped will result in total occlusion of vessel. A rubber vessel loop can also be used, by going around the vessel and both ends are lifted up to kink the vessel to occlude the vessel temporary.

iv) Dissection

Tissue dissection should be performed with the objective of avoiding unnecessary tissue, blood vessel and nerve damage. The two most essential instrument used are a blunt tip dissection scissors like the Metzembaun and an atraumatic forceps. Dissection normally involves a series of tissue lifting with the forcep and teasing and cutting of tissue with the scissors to separate structures and tissue planes. A right angle forceps is useful for vascular isolation and mosquito artery forceps is commonly used for griping of suture ends and tissues. Different atraumatic clamps and graspers are used for specific tissue handling such as Debakey atraumatic or bulldog clamp for vessels; Doyen longitudinally serrated clamp for the intestine; Duval grasper for the lung; Self-retaining bone grasper for the bone; Allis forcep for gastrointestinal tissue; Babcock for the stomach; Satinsky clamp for superior vena cava and Diethrich clamp for the aorta. For microvascular procedure or surgery on rodents, the appropriate instruments to use will include Adson forceps, Jeweller forceps, Iris micro-Bulldog clamp and spring handle microscissors and needle holder and Bremer double approximator clamp. The list of surgical instruments is long and not possible to be fully described within this text.

v) Wound Closure

Surgical wounds are normally closed with double layer tissue suturing. Muscle and subcutaneous tissue can be closed with absorbable suture using uninterrupted over and over suturing procedure. Size of suture depends on load bearing factors and normally for an abdominal wound in large animal like a pig, a size 1 or 0 is used with round tip non-cutting needle. For small animal 3/0 or 4/0 suture should suffice. The skin can be closed with interrupted suturing with monofilament absorbable suture material that comes with triangulated cutting tip needle. Monofilament suture is preferred as it reduced chances of microbial harbourage and wicking of bacteria into the wound. Absorbable suture material is recommended, as it self degrades in a month or so without need for suture removal. Where skin thickness permits, a hidden subcuticular skin closure is used for good skin apposition and no need for suture removal. Undue force should not be used during the knot tying, which will cause tissue strangulation and impedes wound healing. After closure, 1 % tetracycline cream may be applied and covered with a wound dressing (e.g. loban) as a barrier to external microbial contamination.

vi) Wound Drainage

Drains are not usually used in routine animal surgery. Their purpose is to allow the excess body fluids, blood or inflammatory exudate to run out or be sucked away from the wound or body cavities. The two main drains used are open and closed. In open drains, a length of solid or hollow tubing is pulled out of the skin and covered with dressing, sealed and positioned appropriately so that they cannot be interfered with by the animal. In thoracic surgery, closed drainage tubes are attached to a glass or plastic container, the latter usually being evacuated to suck away unwanted fluid via a water seal so as not to compromise the negative pressure in the pleural cavity on inspiration (Fig 3.5.1). The tubing is held to the skin exit with a secured purse string suturing technique.

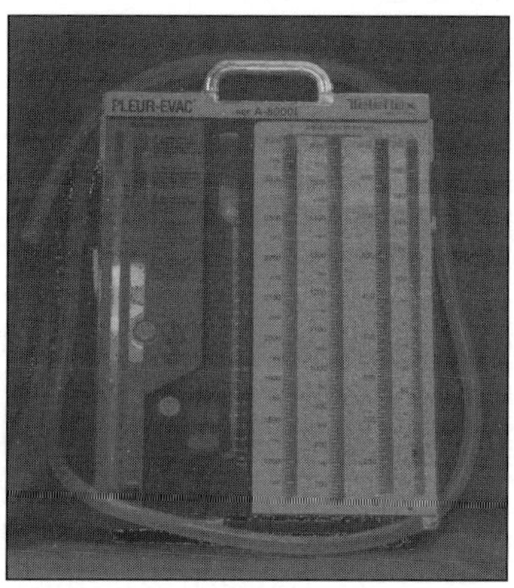

Fig 3.5.1: Underwater seal for chest tube drainage.

CHAPTER **3.6**

ANIMAL INTUBATION

Robert Ng

Intubation is the process of introducing an endotracheal tube into the trachea for volatile anaesthetic inhalation and oxygenation. Different animal species have their own specific level of difficulty in intubation. Alternatives to the use of the endotracheal tube are the use of a nose cone, nasal tube or tracheotomy tube.

Mostly constructed of soft plastic, rubber or silicon, endotracheal tubes vary in diameter from 2.5 mm for rabbits to 9.0 mm for sheep. The most commonly used endotracheal tubes are the cuffed (Fig 3.6.1) and noncuffed Magill tubes. These tubes may contain wire reinforcement that is very helpful in avoiding kinking or twisting especially in smaller animals or in cases where the tube outlet end at the side of the mouth that lies in an awkward position. A plastic stylet inserted into the endotracheal tube is useful for intubation both as a guide and as a rigid support for the soft tube. There is also a special double luminal endotracheal tube commonly used in cardiothoracic surgery (Fig 3.6.3).

Ease of placement of the endotracheal tubes (Fig 3.6.2) into the trachea is significantly enhanced by adequate anaesthetic depth and muscle relaxation of the upper airway and oral area. The use of force during intubation should be avoided. Excessive trauma to the larynx can produce laryngeal oedema, laryngeal spasm or excessive stimulation of the vagus resulting in bradycardia. The use of local anaesthesia such as 10 % lignocaine may be helpful. Care should also be used to avoid excess inflation of the cuff to avoid damage to the trachea.

A straight or curved laryngoscope is used to retract the tongue to expose the epiglottis and visualize the laryngeal opening for intubation. The Miller style blade is often used. A 205 mm blade is used for sheep while a smaller infant McIntosch blade adapts well to rabbits.

Pig

The pig is first chemically restrained with ketamine (15 mg/kg) to prepare for volatile anaesthetic inhalation. The shape of the pig's head is well suited to the use of nose cone for anaesthesia with 5 % isoflurane inhalation. Some anatomical features of the pig are responsible for the difficulty often encountered when attempting intubation besides the fact that the pig's mouth cannot be retracted widely. One such difficulty is that it has a deeply recessed and small larynx that slopes downwards creating a very sharp angle to the tracheal opening.

With the pig in supine position, short lengths of cotton tape are placed around the upper and lower jaw behind the canine teeth to open the pig's mouth as wide as possible. The head and neck are extended, the tongue is drawn forward and a laryngoscope is used to provide an unobstructed view of the laryngeal opening. An appropriately sized endotracheal tube with a plastic stylet is then directed through the cricoid ring into the tracheal opening. When the endotracheal tube has been properly positioned, the cuff is inflated to provide an airtight seal with the tracheal wall and the tube is then connected to the anaesthetic machine.

Alternatively, a DES-patented mechanical mouth gag can be used with the upper and lower blade retracting the jaws by adjusting a geared rachet and the tongue is held in position by a tongue holder. The laryngoscope and endotracheal tube introduction can then be performed easily. The whole process can be performed by a single operator.

Rabbit

Intubating a rabbit can be a difficult procedure because of the animal's large tongue, skin folds in the diastema, limited range of mandibular opening and prominent incisors that obstruct the placement of an endotracheal tube. Several techniques have been devised to simplify the difficult task of endotracheal intubation in rabbits. The most important consideration for rabbit intubation is to ensure that the animal is totally relaxed.

Aided by the use of a small blade laryngoscope, the visual placement of a small endotracheal tube is possible after some practise. To assist in the visualization and tube placement, a 3 mm endoscope can be used in conjunction with a laryngoscope for video image guidance via a camera console.

Blind intubation, with the larynx grasped through the skin and tube placement is guided by feel, is practiced in some animal centers. It is, however, not recommended for the novice as significant trauma is inevitable with bleeding and laryngeal swelling to follow.

A less traumatic method is the use of the otoscope cone and a guide tube (like a small polyprophylene tube) which, may be placed into the trachea. After removal of the otoscope, a 2.5 mm endotracheal tube may be advanced over the guide tube, which is then removed.

Endotracheal intubation is not required in all rabbit anaesthesia. All the techniques suggested above require practice to ensure success without undue delay and injury. However, in certain surgical procedures such as in cardiothoracic surgery, endotracheal intubation is mandatory for intermittent positive pressure ventilation.

Mice and Rat

In rodents, volatile anaesthetic inhalation provides for relatively easy adjustment of depth of anaesthesia and is safer for prolonged or otherwise demanding procedure. A wide variety of anaesthetic induction chambers is available and a simple system can be fashioned from a plastic box with two ports, one for inflowing oxygen and volatile anaesthetic via a vapourizer and the other a scavenging port for used anaesthetic discharge. For anaesthesia maintenance, a nose cone is used that is designed from a wire-reinforced gastrotube with the cone end applied to the animal nose and a side tubing for delivery of gas from an anaesthetic vapourizer. Waste gas flows to a scavenging system via the main tube proximal to the cone end. The internal diameter of the nose cone for rat is 12.5 mm and for mice a size 7 mm is more appropriate.

For rat intubation, one technique involves the use of a 2 mm endoscope where the stem is used as tongue depressor and the 0 ° scope is used to visualize the epiglottis via monitor video imaging for placement of a 20 G plastic cannula with the luer lock mounted end connected to a Harvard ventilator. There are also commercially available rodent intubation kits.

Alternative techniques involve blind and tracheostomy intubation. Both methods involve surgical exposure of the trachea and for blind procedure requires the insertion of a catheter tube through the mouth and depends the finger feel on the exposed trachea to guide the tube into the trachea. This technique requires a high degree of experience. The other method involves surgical separation of paratracheal muscle to expose the trachea for a trocar catheter insertion to be made between the cartilaginous rings. The needle component of the catheter is removed and the indwelling plastic component is directed into the trachea leaving the luer lock mount end for connection to a ventilator. Another way is to make a small transverse incision on the trachea enough for the introduction of a polyprophylene tube, which is passed cranially to exit out of the animal mouth. The caudal end of the tube is then inserted back into the trachea. The trachea and skin wound is suture closed. The external end of the tube is inserted with a needle and the luer mount end connected to the ventilator.

Nonhuman Primates

Intubation of nonhuman primate is not as difficult as for similarly sized animals like the rabbits. A curved baby Miller laryngoscope can be used. A laryngoscope fashioned from a battery operated penlight can also be used. The pen light has an attached spatula which is slightly curved and has a rounded end that protrudes 2 cm above the light source. Being of slim structure, it is also easier to access the mouth, which is opened by depressing the cheek area near the mandibular joint. The animal head should not be extended but maintained at slight flex to the neck with the anaesthetized animal lying on a supine position. A 2.5 mm cuffed endotracheal tube is used for the intubation. For minor procedures, a mouth muzzle mask is used to deliver volatile anaesthetic from the anaesthetic machine.

Sheep

Adult sheep have a large tracheal opening for the insertion of a size 9 or 10 mm endotracheal tube aided by a long blade laryngoscope. As soon as the endotracheal tube has been inserted, it is advisable to insert a gastric tube (example, a 15 mm diameter plastic tube) into the oesophagus. Sheep in particular are prone to rumental lympany with resultant obstruction of the surgical field, respiratory embarrassment due to pressure exerted on the diaphragm and depressed cardiac output due to depressed venous return. If gas still accumulates despite the gastric tube insertion, a 14 gauge needle should be inserted into the swollen rumen before it becomes tense. The head of the animal is inclined downwards to allow saliva and rumenal fluid to drain away from the mouth into a container bag.

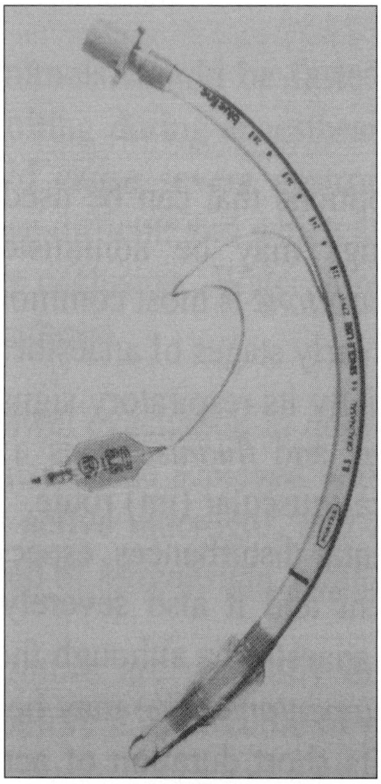

Fig 3.6.1: Standard cuffed endotracheal tube.

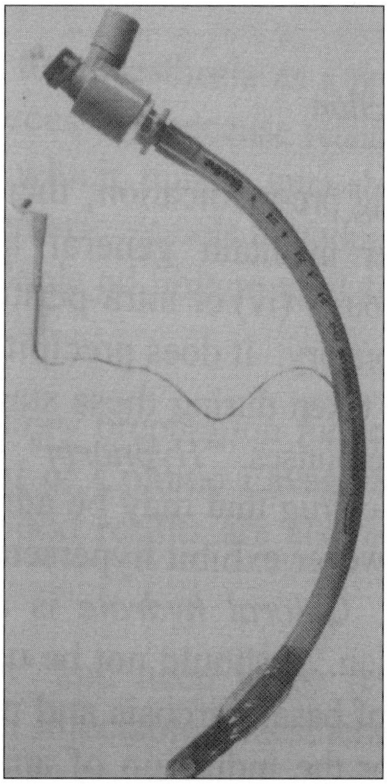

Fig 3.6.2: Endotracheal tube for bronchoscopy.

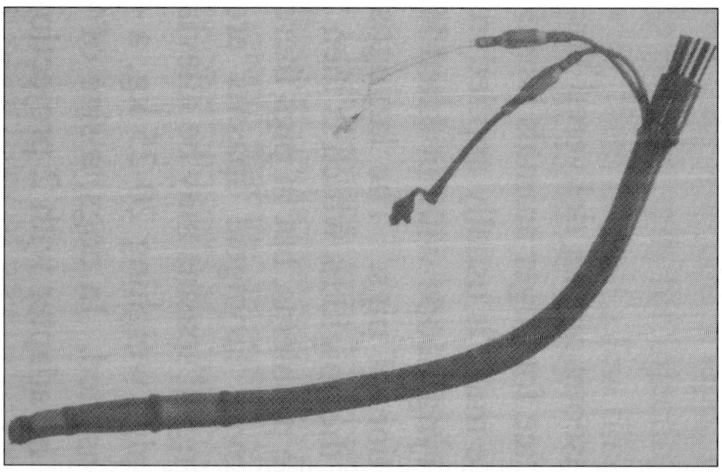

Fig 3.6.3: Double lumenal endotracheal tube for cardiothoracic surgery.

CHAPTER **3.7**

ANAESTHESIA AND MAINTENANCE OF HOMEOSTASIS

Robert Ng

General anaesthesia is a state of general depression of the central nervous system (CNS) and involves hypnosis, analgesia, suppression of reflex activity and relaxation of voluntary muscles. Successful anaesthesia is intimately linked to good maintenance of homeostasis. The physiologic condition of the anaesthetized animal is closely monitored. This includes observing the mucuous membrane colour, capillary refill time, ascultation, blood pressure, respiratory volume and rate, and body temperature. The progressive steps in animal anaesthesia will involve restraining (physical or chemically induced), sedation (induced drowsiness and decreased motor activity), induction of anaesthesia (induction for intubation purposes) and maintenance anaesthesia (sustained level of surgical anaesthesia plane).

Anaesthetic Agent

Agents used can be broadly classified as sedatives (xylazine), hypnotics (chloral hydrate), tranquilizers (acepromazine), muscle relaxant (diazepam), anticonvulsant (phenobarbitone), dissociative anaesthetics (ketamine), neuroleptonalgesic (fentanyl-fluanisone), and volatile agent (isoflurane). For animal anaesthesia, it is useful to focus on the few agents that the research institution is familiar with.

- Anticholinergic drug: atrophine sulphate is used to diminish salivary and bronchial secretions to protect the heart from vagal inhibition. It should be routinely used when ketamine is used. During general anaesthesia, glycopyrrolate can be used for a longer duration of action.

- Dissociative anaesthetic: ketamine is the agent of choice for most experimental animals and is best used in combination with diazepam, xylazine or acepromazine.
- Hypnotic agents: xylazine is the sedative of choice at DES for sheep and goat. Its pronounced muscle relaxant and analgesic properties render it useful in combination with ketamine for rabbit.
- Tranquilizer agent: diazepam (valium) is an extremely useful agent for rodents and most experimental animals when used in combination with dissociative agent, ketamine, especially for laparoscopic surgeries.
- Volatile anaesthetic: isoflurane, is an indispensable agent for inhalation anaesthesia delivered via a vapourizer with oxygen.

Certain anaesthetic agents are prohibited for use such as chloral hydrate due to its severe depressive action on the respiratory and cardiovascular systems. Ether (diethyl ether) use as a volatile agent is not recommended due to its flammable property. Another volatile agent halothane is also not recommended because 25 % of what is inhaled need to be metabolized in the liver as compared to isoflurane (1 %). Pancuronium, a neuromuscular blocker, paralyses the skeletal muscle and hence, many signs of anaesthetic depth are masked because of this paralysis and it is therefore not normally recommended.

Pre-operative Preparation

For short procedures, inhalational anaesthesia with 2 % isoflurane is preferred especially for rodents as it facilitates speedy recovery and involves minimal animal handling. For long surgical procedures, ketamine is the choice sedative/anaesthetic agent and is administered intramuscularly or intravenously after physical restraining of animal. It is often used in combination with other agent like atrophine sulphate (anticholinergic), diazepam (tranquilizer) or xylazine (a hypnotic). Anaesthesia is then induced with inhalation of isoflurane at 3 to 5 %, delivered through a mouth-breathing muzzle. A totally relaxed animal will ease the intubation procedure; otherwise laryngeal spasm can make the process complicated though it can be relieved with 10 % lignocaine spray application.

The animal is intubated with an appropriately sized endotracheal tube. Cuffed endotracheal tube is normally used for Intermittent Positive Pressure Ventilation (IPPV). It is also recommended for spontaneous ventilation as it allows convenient conversion to IPPV mode in the event of respiratory complication. For bronchoscopy, the endotracheal tube comes with an extra port opening for insertion of endoscope. For cardiothoracic surgery, a double lumen endotracheal tube is sometimes used to deflate one lung to provide sufficient chest space for surgery.

The intubated animal is hooked up to the anaesthetic machine (Fig 3.7.1), which is preset to appropriate tidal volume (10 to 15 ml/kg weight), airway pressure (20 to 25 mmHg), breathing frequency (12 to 15/minute), higher rate for rodents, and isoflurane concentration (1 to 3 %). The animal is hooked up to monitor ECG pattern, $Sp0_2$ and rectal temperature. The animal is checked for response to stimuli to ascertain depth of surgical anaesthesia using pedal reflex, ear pinch, blink or palpebral reflex and jaw opening resistance. General anaesthesia is normally prescribed for animals as it is difficult to restrain animal with topical anaesthesia with 10 % lignocaine. As intravenous (IV) infusion line is established for fluid maintenance, replacement of blood loss and for drug administration if required, via cannulation of ear, jugular, cephalic or the saphenous vein. The bag of IV fluid container is sometimes placed within a pressure bag for assisted pressurized delivery when the cannulated vein is small and exerts significant flow impedance. An arterial line, for invasive blood pressure (IBP) monitoring, is established under sterile condition and this involves doing a cut-down to access the carotid, tibial or femoral artery and the line linked to a pressure transducer. For cardiovascular surgery, a syringe pump set up is required for infusion of propofol (0.5 ml/kg) to provide alternative anaesthetic support when the ventilator is shut down once vascular bypass through the extracoporal heart lung machine is established. A defibrillator should also be on standby for emergency application. A manometric central venous pressure line is sometimes required for monitoring venous return especially in major operation involving the need for vascular clamping. A handheld blood gas analyser, I-Stat, comes in useful for the measurement of electrolytes, pO_2, pCO_2 and pH during major surgeries as in organ transplant or cardiothoracic surgeries. To monitor expired air, a CO_2 and isoflurane concentration monitor (Fig 3.7.2) can be used.

Intra-operative Anaesthetic Management

The principal objective will be to maintain physiological stability during anaesthesia, and include close monitoring of physiological parameters and the maintenance of anaesthetic performance.

A. *Respiratory Complication*

Respiratory complications can result from inadequate ventilation caused by depression of respiratory function. This may be due to hypoxia, narcotic overdose or airway obstruction secondary to paralysis of anatomical soft structures or by excessive saliva, bronchial secretions or blood. Signs of respiratory complications include respiratory arrest (apnoea), hyperventilation (hypernoea), hypoventilation (hyponoea), gasping respiration and accelerated respiration rate (tachypnoea).

If apnoea is observed, artificial respiration should be initiated by squeezing the rebreathing bag or pressing on the chest rhythmically to encourage movement of gas in the lung. Hypoventilation due to poor ventilation is potentially dangerous since CO_2 elimination is inefficient and immediate conversion to IPPV is recommended. Airway obstruction may result from laryngospasm, bronchospasm or kinking of the endotracheal tube. Relieving of bronchospasm can be achieved with 10 % lignocaine spraying of the larynx.

Respiratory problems can be largely avoided if animal is placed on IPPV and when endotracheal insertion is trouble free. Furthermore, where practical, a gastric tube should be inserted to further protect the airway and to prevent gastric distension which exerts pressure on the diaphragm.

B. *Cardiovascular Failure or "Shock"*

Acute cardiovascular failure causes failure of blood flow resulting in inadequate tissue perfusion and hence cellular hypoxia. This is caused mainly by haemorrhage and depressed cardiac function. Clinical symptoms include pallor of the skin, mucuous membranes, weak pulse, and decreased circulating blood volume. Blood volume can be expanded rapidly with Hartman's solution to improve cardiac output or transfusion with blood or plasma expander like gelofusin to prevent interstitial edema. Cardiac output can also be increased with isoprenaline, doxapram or dopamine.

C. *Cardiac Arrest*

Ventricular fibrillation is a terminal event. To be successful, emergency therapy must restore circulation to brain tissue within three minutes with stimulation of the myocardium by adrenaline or the use of IV injection of calcium gluconate to enhance contractility of the dilated and flaccid heart. Ventricular fibrillation can be reversed with an electric defibrillator application.

D. *Hypothermia*

Anaesthetics and most central nervous system (CNS) depressants override the thermoregulatory centre so that the animal is unable to respond to a fall in body temperature in such ways as shivering or increased insulation by piloerection. It is of particular importance to note that the maintenance of normal body temperature will minimize cardiovascular and respiratory disturbances caused by anaesthetic agents. A few ways to conserve body heat is for the operating table to have heated pads and IV infusion should be warmed to 38 °C. Covering animals with heavy blankets and warming with overhead heater will also help.

E. *Acid Base Balance*

Acid base inbalance could either be a result of respiratory pattern changes or metabolic disturbances. Metabolic accidents may be caused by loss of sodium bicarbonate due to diminished renal excretion of acid, accumulation of ketonic acid in the body and release of acidic products, such as lactic acid into the circulation from tissues subjected to ischaemia, as in organ replantation or surgeries where there is disruption of the vascular flow due to vascular occlusion from cross clamping of large vessel.

Respiratory acidosis is mainly due to poor lung ventilation resulting in retention of CO_2 and hence an increase in carbonic acid (H_2CO_3) concentration. Acidic condition can be reversed with IV infusion of alkalinizing agent, sodium bicarbonate, which is conveniently supplied as an 8.4 % solution so that 1 ml will contain 1 mg of $NaHCO_3$. A dosage of 5.0 mg/kg $NaHCO_3$ can be given safely to subjects. Metabolic alkalosis may result from excess loss of acid through vomiting or by excessive loss of potassium ions via the kidneys associated with overdosing with adrenal corticosteroids. Respiratory alkalosis is contributed by hyperventilation, which is only of importance when animal is on IPPV.

F. *Diuretics*

It is important that the kidneys are given every opportunity to maintain adequate function during surgery and anaesthesia so that anaesthetic agent and their metabolites are eliminated quickly. Adequate renal arterial pressure and a fluid load are the best insurances against renal failure but two specific agents, namely 10 %

mannitol (500 to 1000 mg/kg iv) or furosemide (Lasix) at 2 to 4 mg/kg iv can be administered.

G. *Anaesthetic Recovery*

Before reversal, it is important to ensure that animal is physiologically stable. If heparin is given during surgery, it needs to be neutralized with protamine sulphate at 2 mg/kg for every 1.5 mg/kg of heparin administered.

For animals anaesthetized under IPPV mode, forced oxygen respiration is slowly weaned off starting with 50 % mixture of air and oxygen. Ventilation is switched off to spontaneous mode with rebreathing bag application. SpO_2 reading is used as reference to induce slight hypoxia by repeated lowering to 80 % to encourage onset of spontaneous respiration of the animals.

Recovery may differ in its time course but it follows a similar pattern. The palpebral and respiratory reflexes return in that order. Muscular trembling may then start and the chewing action will follow. Complete recovery with the animal fully ambulatory, eating and drinking may take minutes or up to 48 hours depending on the drugs, species of animal and surgical insult inflicted.

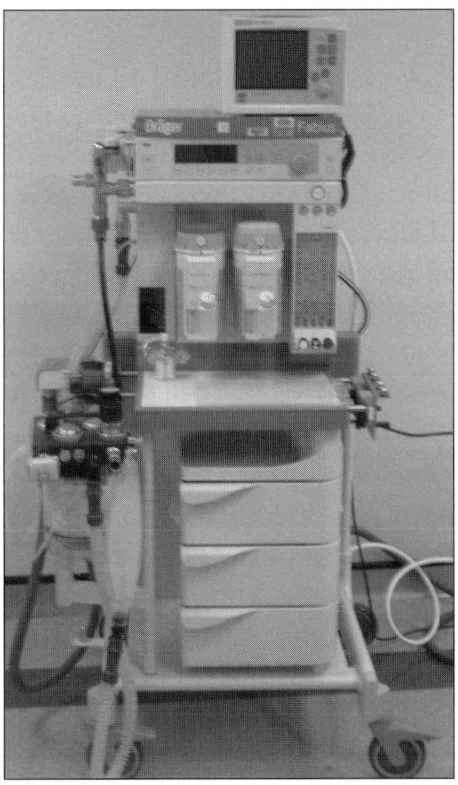

Fig 3.7.1: Anaesthesia machine.

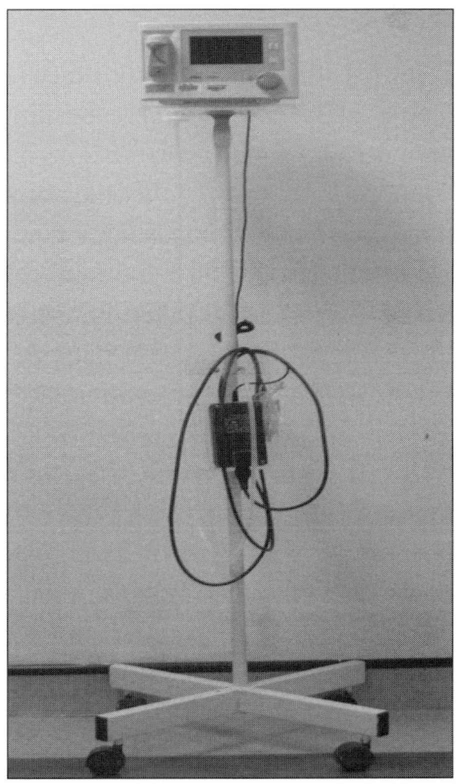

Fig 3.7.2: CO_2 and halothane concentration monitor of expired air.

CHAPTER 3.8

ANIMAL EUTHANASIA

Darvi Sergio

The term euthanasia actually means "good death". It is derived from the Greek term *eu* meaning good and *thanatos* meaning death. A good death is one in which the animal experiences no pain, no fear and no other significant stress before dying. This means that the death occurs instantly or that the animal is first rendered fully unconscious through painless methods. Humane euthanasia is a component of most animal research projects.

Humaneness is an important responsibility of principal investigators, veterinarians, veterinary staff and all other human beings entrusted with the care and use of animals. If an animal's life is to be taken it must be done humanely with the highest degree of respect for the animal, and with an emphasis on making the death as painless and distress free as possible.

Most investigators who use animals must eventually face the need to bring about the animals' death. Animals are normally euthanized at the end of the study for sample collection or post-mortem examination and their tissues processed for analysis. Some animals are euthanized because their scientific usefulness has ended. Animals may also be euthanized if they are experiencing pain or distress.

The obligation to provide laboratory animals with a stress-free and painless death is mandated by both good experimental design and humane considerations. No euthanasia method is likely to be carried out unless it is compatible with the investigators' research needs and is aesthetically acceptable to the persons who are tasked to execute it. In addition, a good euthanasia must be safe to people.

Moreover, the obligation is backed by law and supported by guidelines established by the *National Advisory Committee for Laboratory Animal Research* (*NACLAR*).

In practical terms, every investigator should understand that research project applications must include the experimental endpoints and euthanasia methods. This should conform to the *NACLAR* guidelines for euthanasia and must have approval by the *IACUC*.

The investigator must choose a method of euthanasia that does not interfere with the scientific objectives of the project. If the animal's tissues are required for analysis, then the euthanasia method selected should not render the tissue unuseable. Investigators need to familiarize themselves with the detailed effects of various euthanasia methods before they can make an intelligent choice.

Criteria for Euthanasia

Euthanasia must be considered if animals demonstrate the conditions listed below, whether the animal has been manipulated/handled or not. Fulfillment of one criterion can constitute grounds for euthanasia. **Exceptions** are permitted only if approved by the IACUC as part of the protocol review process. Some of these criteria are listed as follows:

1. *Weight loss*
   Weight loss of 20 to 25 % (depending on body condition, weight recorded at time of arrival, and age: growing animals may not lose weight, but may not gain normally if ill). If body weight is not measured, weight loss may be characterized by cachexic muscle wasting. This may be detected by body condition scoring.

2. *Inappetence*
   Complete anorexia for 24 hours in small rodents, up to 5 days in large animals; partial anorexia (less than 50 % of caloric requirement) for 3 days in small rodents, 7 days in large animals.

3. *Weakness/Inability to obtain feed or water*
   Inability or extreme reluctance to stand which persists for 24 hours, assuming that the animal has recovered from anaesthesia.

4. *Moribund state*
   Depression coupled with body temperature below 37.2 °C or nonresponsive to stimulation assuming that the animal has recovered from anaesthesia.

5. *Infection*
   Infection involving any organ system (either overt, or indicated by increased body temperature or WBC parameters), which fails to respond to antibiotic therapy within any appropriate time and is accompanied by systemic signs of illness.

6. *Signs of severe organ system dysfunction non-responsive to treatment, or with a poor prognosis as determined by a veterinarian, for example:*

Respiratory:	Dyspnoea, cyanosis.
Cardiovascular:	Blood loss or anaemia resulting in haematocrit below 20 %; one transfusion may be performed.
Gastrointestinal:	Severe vomiting or diarrhoea, obstruction, intussusception; peritonitis; evisceration (immediate euthanasia required); only one major surgical procedure (involving entry of abdomen or thorax) may be performed per animal unless indicated on approved protocol. Therefore, major surgery intended to correct complications may not be permitted without prior approval from the IACUC or the attending veterinarian in emergency situations. If approval is not granted, euthanasia must be performed.
Urogenital:	Renal failure characterized by elevated BUN, creatinine or uroperitoneum.
Nervous:	CNS depression, seizures, paralysis of one or more extremities; pain unresponsive to analgesic therapy.
Musculoskeletal:	Muscle damage, bone injury, locomotor deficits, etc. resulting in inability to use the limb, unless anticipated as part of the study.
Integumentary:	Non-healing wounds, repeated self-trauma, second or third degree heating pad burns of an extensive nature.

Methods of Euthanasia

The choice of a method of euthanasia depends on species, age, availability of restraint, skill of the individuals performing euthanasia and other considerations. The method of euthanasia chosen by the investigator must be consistent with the research goals and objectives.

IACUC reviews the approved methods of euthanasia based on the following:

- ability to induce loss of consciousness and death without causing pain, distress and anxiety, or apprehension;
- time required to induce loss of consciousness;
- reliability and irreversibility;
- safety of personnel caring out the euthanasia, as well as emotional effect on observers and operators;
- compatibility with subsequent evaluation, examination or use of tissue;
- drug availability and human abuse potential;
- compatibility with species, age and health status.

Table 3.8.1 lists some of the acceptable methods for euthanasia while other conditionally acceptable methods that require IACUC approval are given in Table 3.8.2.

Table 3.8.1: List of acceptable euthanasia methods for laboratory animals

Method	**Species of animals**
Barbiturates	Most species
Carbon dioxide (bottled gas only)	Mice and rats
Inhalant anaesthetics	Most species
Microwave irradiation	Mice and rats
Tricaine methane sulfate (TMS, MS222)	Fish and amphibians
Benzocaine hydrochloride	Fish and amphibians
Captive penetrating bolt	Horse, ruminant and swine

Table 3.8.2: List of conditionally acceptable methods for euthanasia

Method	**Species of animals**
Cervical dislocation	Birds, rodents and rabbits
Decapitation	Birds, rodents and some other species
Pithing	Some ecthotherms
Various pharmacological and physical methods	Most species

There are also methods unacceptable for use in animal euthanasia. These are listed below:

- chloral hydrate, chloroform and cyanide
- decompression
- neuromuscular blockers
- various pharmacological and physical methods
- dry ice generated CO_2.

Agents and methods of euthanasia by species (*JAVMA*, Vol. 218, No. 5, March 1, 2001. 2000 Report of the AVMA Panel on Euthanasia).

Species	**Acceptable methods***	**Conditionally acceptable (requires IACUC approval of scientific justification)†**
Cats	Barbiturates, inhalant anesthetics, CO_2, CO, potassium chloride in conjunction with general anesthesia	N_2, AR
Dogs	Barbiturates, inhalant anesthetics, CO_2 (mink require high concentrations for euthanasia without supplemental agents), CO, potassium chloride in conjunction with general anesthesia	N_2, Ar, penetrating captive bolt, electrocution
Nonhuman primates	Barbiturates, inhalant anesthetics, CO_2, CO, potassium chloride in conjunction with general anesthesia	N_2, Ar, cervical dislocation (< 1 kg), decapitation, penetrating captive bolt
Rabbits	Barbiturates, inhalant anesthetics (in appropriate species), CO_2 (in appropriate species)	Penetrating captive bolt, gunshot, decapitation and pithing, stunning and decapitation
Rodents and other small mammals	Barbiturates, potassium chloride in conjunction with general anesthesia, penetrating captive bolt	Chloral hydrate (IV, after sedation), gunshot, electrocution
Ruminants	Barbiturates, CO_2, potassium chloride in conjunction with general anesthesia, penetrating captive bolt	Inhalant anesthetics, CO, chloral hydrate (IV, after sedation), gunshot, electrocution, blow to the head (< 3 weeks of age) N_2, Ar, penetrating captive bolt, gunshot
Swine	Barbiturates, inhalant anesthetics, CO_2, CO, potassium chloride in conjunction with general anesthesia	CO_2, CO, N_2, Ar, penetrating captive bolt, gunshot

*Acceptable methods are those that consistently produce a humane death when used as the sole means of euthanasia.

†Conditionally acceptable methods are those that by the nature of the technique or because of greater potential for operator error or safety hazards might not consistently produce humane death or are methods not well documented in the scientific literature.

Acceptable agents and methods of euthanasia — characteristics and modes of action (*JAVMA*, Vol. 218, No. 5, March 1, 2001. 2000 Report of the AVMA Panel on Euthanasia).

Agents	Classification	Mode of action	Rapidity	Ease of performance	Safety for personnel	Species suitability	Efficacy and comments
Barbiturates	Hypoxia attributable to depression of vital centers	Direct depression of cerebral cortex, subcortical structures, and vital centers; direct depression of heart muscle	Rapid onset of anesthesia	Animal must be restrained; personnel must be skilled to perform IV injection	Safe except human abuse potential; DEA-controlled substance	Most species	Highly effective when appropriately administered; acceptable IP in small animals and IV
Carbon dioxide (bottled gas only)	Hypoxia attributable to depression of vital centers	Direct depression of cerebral cortex, subcortical structures, and vital centers; direct depression of heart muscle	Moderately rapid	Used in closed container	Minimal hazard	Small laboratory animals, birds, cats, small dogs, rabbits, mink (high concentrations required), zoo animals, amphibians, fish, some reptiles, swine	Effective, but time required may be prolonged in immature and neonatal animals
Carbon monoxide (bottled gas only)	Hypoxia	Combines with hemoglobin, preventing its combination with oxygen	Moderate onset time, but insidious so animal is unaware of onset	Requires appropriately maintained equipment	Extremely hazardous, toxic, and difficult to detect	Most small species including dogs, cats, rodents, mink, chinchillas, birds, reptiles, amphibians, zoo animals, rabbits	Effective; acceptable only when equipment is properly designed and operated
Inhalant anesthetics	Hypoxia attributable to depression of vital centers	Direct depression of cerebral cortex, subcortical structures, and vital centers	Moderately rapid onset of anesthesia, excitation may develop during induction	Easily performed with closed container; can be administered to large animals by means of a mask	Must be properly scavenged or vented to minimize exposure to personnel	Some amphibians, birds, cats, dogs, furbearing animals, rabbits, some reptiles, rodents and other small mammals, zoo animals, fish, freeranging wildlife	Highly effective provided that subject is sufficiently exposed; either is conditionally acceptable
Microwave irradiation	Brain enzyme inactivation	Direct inactivation of brain enzymes by rapid heating of brain	Very rapid	Requires training and highly specialized equipment	Safe	Mice, rats	Highly effective for special needs
Penetrating captive bolt	Physical damage to brain	Direct concussion of brain tissue	Rapid	Requires skill, adequate restraint, and proper placement of captive bolt	Safe	Horses, ruminants, swine	Instant loss of consciousness, but motor activity may continue
Potassium chloride (intracardially or intravenously in conjunction with general anesthesia only)	Hypoxia	Direct depression of cerebral cortex, subcortical structures, and vital centers secondary to cardiac arrest	Rapid	Requires training and specialized equipment for remote injection anesthesia, and ability to give IV injection of potassium chloride	Anesthetics may be hazardous with accidental human exposure	Most species	Highly effective, some clonic muscle spasms may be observed

Conditionally acceptable agents and methods of euthanasia — characteristics and modes of action (*JAVMA*, Vol. 218, No. 5, March 1, 2001. 2000 Report of the AVMA Panel on Euthanasia).

Agents	Classification	Mode of action	Rapidity	Ease of performance	Safety for personnel	Species suitability	Efficacy and comments
Carbon dioxide (bottled gas only)	Hypoxia due to depression of vital centers	Direct depression of cerebral cortex, subcortical structures and vital centers; direct depression of heart muscle	Moderately rapid	Used in closed container	Minimal hazard	Nonhuman primates, freeranging wildlife	Effective, but time required may be prolonged in immature and neonatal animals
Cervical dislocation	Hypoxia due to disruption of vital centers	Direct depression of brain	Moderately rapid	Requires training and skill	Safe	Poultry, birds, laboratory mice, rats (< 200 g), rabbits (< 1 kg)	Irreversible; violent muscle contractions can occur after cervical dislocation
Decapitation	Hypoxia due to disruption of vital centers	Direct depression of brain	Rapid	Requires training and skill	Guillotine poses potential employee injury hazard	Laboratory rodents; small rabbits; birds; some fish, amphibians, and reptiles (latter 3 with pithing)	Irreversible; violent muscle contraction can occur after decapitation
Electrocution	Hypoxia	Direct depression of brain and cardiac fibrillation	Can be rapid	Not easily performed in all instances	Hazardous to personnel	Used primarily in sheep, swine, foxes, mink (with cervical dislocation), ruminants, animals > 5 kg	Violent muscle contractions occur at same time as loss of consciousness
Inhalant anesthetics	Hypoxia due to depression of vital center	Direct depression of cerebral cortex, subcortical structures, and vital centers	Moderately rapid onset of anesthesia; excitation may develop during induction	Easily performed with closed container; can be administered to large animals by means of a mask	Must be properly scavenged or vented to minimize exposure to personnel; ether has explosive potential and exposure to ether may be stressful	Nonhuman primates, swine; ether is conditionally acceptable for rodents and small mammals; methoxyflurane is conditionally acceptable for rodents and small mammals.	Highly effective provided that subject is sufficiently exposed
Penetrating captive bolt	Physical damage to brain	Direct concussion of brain tissue	Rapid	Requires skill, adequate restraint and proper placement of captive bolt	Safe	Dogs, rabbits, zoo animals, reptiles, amphibians, free-ranging wildlife	Instant loss of consciousness but motor activity may continue
Pithing	Hypoxia due to disruption of vital centers, physical damage to brain	Trauma of brain and spinal cord tissue	Rapid	Easily performed but requires skill	Safe	Some ectotherms	Effective, but death not immediate unless brain and spinal cord are pithed
Thoracic compression	Hypoxia and cardiac arrest	Physical interference with cardiac and respiratory function	Moderately rapid	Requires training	Safe	Small- to medium-sized free-ranging birds	Apparently effective

Some unacceptable agents and methods of euthanasia (*JAVMA*, Vol. 218, No. 5, March 1, 2001. 2000 Report of the AVMA Panel on Euthanasia).

Agents	**Comments**
Air embolism	Air embolism may be accompanied by convulsions, opisthotonos, and vocalization. If used, it should be done only in anesthetized animals.
Blow to the head	Unacceptable for most species.
Chloral hydrate	Unacceptable in dogs, cats, and small mammals.
Chloroform	Chloroform is a known hepatotoxin and suspected carcinogen and, therefore, is extremely hazardous to personnel.
Cyanide	Cyanide poses an extreme danger to personnel and the manner of death is aesthetically objectionable.
Decompression	Decompression is unacceptable for euthanasia because of numerous disadvantages. (1) Many chambers are designed to produce decompression at a rate 15 to 60 times faster than that recommended as optimum for animals, resulting in pain and distress attributable to expanding gases trapped in body cavities. (2) Immature animals are tolerant of hypoxia, and longer periods of decompression are required before respiration ceases. (3) Accidental recompression, with recovery of injured animals, can occur. (4) Bleeding, vomiting, convulsions, urination, and defecation, which are aesthetically unpleasant, may develop in unconscious animals.
Drowning	Drowning is not a means of euthanasia and is inhumane.
Exsanguination	Because of the anxiety associated with extreme hypovolemia, exsanguinations should be done only in sedated, stunned, or anesthetized animals.
Formalin	Direct immersion of an animal into formalin, as a means of euthanasia, is inhumane.
Household products and solvents	Acetone, quaternary compounds (including CCl_4), laxatives, clove oil, dimethylketone, quaternary ammonium products*, antacids, and other commercial and household products or solvents are not acceptable agents for euthanasia.
Hypothermia	Hypothermia is not an appropriate method of euthanasia.
Neuromuscular blocking agents (nicotine, magnesium sulafte, potassium chloride, all curariform agents)	When used alone, these drugs all cause respiratory arrest before loss of consciousness, so the animal may perceive pain and distress after it is immobilized.
Rapid freezing	Rapid freezing as a sole means of euthanasia is not considered to be humane. If used, animals should be anesthetized prior to freezing.
Strychnine	Strychnine causes violent convulsions and painful muscle contractions.
Stunning	Stunning may render an animal unconscious, but it is not a method of euthanasia (except for neonatal animals with thin craniums). If used, it must be immediately followed by a method that ensures death.

CHAPTER **3.9**

RODENT SENTINEL PROGRAMME

Peik Khin Tan

Many animal research facilities design rodent sentinel programmes to effectively detect selected infectious pathogens in time to allow for proper evaluation and control of disease spread. There are two main objectives of implementing sentinel rodents. Firstly, it is to rapidly detect an infectious disease outbreak within existing colonies and secondly, to evaluate and monitor the health status of mice and rats from noncommercial vendors/suppliers to prevent, detect and control the presence of specific infectious pathogens which may adversely affect animal health and/or influence research protocols. Sentinel animals cannot detect all pathogens, but may serve as indicators of adherence to and effectiveness of barrier systems, containment areas, and preventive practices such as cage changing, animal transport and procurement, use of protective equipment and introduction of biological products.

There are over 30 pathogens of rodents that can cause subclinical to clinical infections. Infections with these agents may affect the experimental results. Although commercial vendors have been providing pathogen free rodents, there are still pathogens commonly encountered in the research setting. The agents chosen for screening are those more commonly encountered in laboratory rodents and are primary rodent pathogens, highly contagious and affect research results. Some of these pathogens are potentially zoonotic.

Sentinel animals used for health surveillance are introduced into the existing animals' colony, housed in identical type of caging as the principal colony, and are placed systematically throughout the room. In DES, these sentinel animals are designated and labeled with ear tag and green colored cage cards. The number of sentinels to be tested depends on the total number of animals housed in the principal colony. Moreover, the number of sentinel animals per cage, per rack and per room depends on the health status of existing colony, the source of vendors/suppliers and the budget of the facility. Frequency of testing sentinel animals varies from one research facility to another.

Selection of Sentinel Animal

Outbred, inbred or hybrid rodents can be used as sentinels. Both have advantages and disadvantages. Outbred stocks are cheaper and have more robust immune responses. Inbred strains are expensive, have strain specific infectious disease susceptibilities and have non-infectious disease predilections that limit their life span. Depending on parental strain characteristics, hybrids are more expensive. Immunodeficient mice (such as SCID or nu/nu including heterozygote immunodeficient mice) should not be used as sentinel animals. Generally, immunocompetent, outbred animals are susceptible to disease and develop strong antibody responses. Therefore, Swiss Webster or other outbred mice from Swiss background and outbred, Sprague-Dawley rats are typically used as sentinels.

Age and Sex of Sentinel

Sentinel rodents should be at least eight weeks of age when placed in the room to ensure the absence of maternal antibodies which could confound results. They must also be maintained in same sex groups. Female littermates are recommended over males due to decreased aggressiveness and fighting compared to males when housed together. When housed in rooms where hormonal influences may affect research objectives, the sex of the sentinels must be considered.

Number of Sentinels and Identification

There should be one or two cages with two or three sentinel cohoused mice per rack or one-two co-housed sentinel rats per full standard rack. Generally, two sentinels are used so that one animal can be tested and serum from the second animal saved in the event of an equivocal or positive result from the first animal. Multiple sentinel cages per room maximizes ability to detect infection.

If the rodent colony is housed using open-top shoeboxes and have at least 100 rodents, the following formula can be used. The number of animals chosen to be screened can be determined using a statistical formula. This formula has a number of assumptions which include: a population of at least 100, the pathogens is randomly dispersed throughout the colony, there is no sex predilection, the percentage of infected animals should be known, and the tests used for screen are 100 % sensitive and specific. The formula used is:

$$\frac{\textbf{Log 0.05}}{\textbf{Log } \boldsymbol{N}} = \textbf{Number to be sampled}$$

where N = the percentage of uninfected animals and
Log 0.05 indicates a 95 % confidence level.

Using this formula, Table 3.9.1 highlights the sample size required to detect at least one positive animal with a 95 % confidence.

Table 3.9.1: Expected incidence of infection in the population and corresponding sample size

Expected incidence of infection in the population (%)	Sample size*
90	2
80	2
70	3
60	4
50	5
40	6
30	9
20	14
10	29
1	298

Most viral diseases have an incidence of infection of 30 to 40 % in conventional housing situations, meaning six to nine animals per 100 should be screened routinely. Unfortunately, many of the assumption are unknown, inappropriate or significantly changed because of husbandry practices often resulting in a requirement for increased numbers of animals to be screened to ensure the same confidence detection.

Obviously it is impractical financially to screen with this degree of rigour. However, to improve ability to detect pathogens and provide reasonable assurances to detect a pathogen, soiled (that is, dirty, used) bedding from colony animals is placed in the boxes of sentinel animals with each bedding change. Using this method, two mice per 100 boxes are sufficient for a sentinel programme. Sentinels cages must be labelled properly (that is, strain, source, date of sentinel placement, date of birth/age and sex) and the cage card should indicate that the sentinel rodents are not to be used for any research purpose.

Caging, Husbandry, Handling and Exposure

Sentinel rodents must be housed in similar fashion to the standard housing of the room. For example, if the colony is housed in open-top rodent cages, sentinels should also be housed in open-top rodent cages (that is, without filter bonnets or micro-isolator tops). If the colonies are housed within closed microisolators or ventilated cages, sentinels should also be housed similarly. The sentinels must be provided with the same feed as the experimental rodents in the room. Handling of sentinel animals like cage changing should be performed only after all other animals have had their cages changed. This is especially important if the sentinels are in a room housing immunocompromised or irreplaceable (such as, transgenic) animals.

Once a group of sentinel rodents has been placed in a room on a particular rack, they must remain with that same group of animals. Sentinels should never be moved from room to room and/or rack-to-rack between different populations/sources of rodents in order to be traceable if there is any positive result. Sentinel mice will have the same weekly cage change schedule as experimental/breeding animal's cage change. Sentinel cages are to be changed last in the room to maximize chances of infectious agent transfer into research cages.

Every time the cages are changed in the room, small samples of soiled, dirty bedding (that is, one to two teaspoons per cage) should be removed from every experimental rodent cage and placed in the sentinel rodent cages. If a large number of cages need to be sampled into a single sentinel cage, a rotation system can be used where 10 to 15 cages are sampled per week. If a rotation system is used, it is critical that all the cages within the colony are sampled within two months to allow exposure and seroconversion to occur in the sentinel animals before they are tested. The composite sample should always include faeces and urine. Sentinel animals should be maintained in 100 % previously utilized, soiled bedding as described. This ensures the sentinel rodents are exposed to any and all potential pathogens in the room. Thus, sentinel rodent bedding will never look as clean as bedding in other cages.

By exposing them to dirty bedding from different cages, the sentinel animals will pick up infectious agents indirectly from bedding soiled with urine and feces. However, some viruses and agents are poorly transmitted through contaminated bedding such as Sendai virus and some external parasites. In such a scenario, it might be necessary to house sentinel together with experimental animals to detect possible pathogens. If contact sentinels are used, there will be a high risk of cross contamination. Therefore, one should weigh the pros and cons of the use of contact sentinels compared to indirect sentinels.

Sentinel rodent cages are often should be placed on the lowest shelf, bottom right corner of the rack under the experimental rodents for system standardization. This ensures any dust, dander, aerosols, microorganisms and parasites, drift down into the sentinel cages, enhancing exposure to potential pathogens.

Frequency of Testing and Replacement of Sentinels

Normal animals seroconvert within two to three weeks of exposure to a virus. Therefore, testing before three weeks of exposure could result in false-negative results. Quarterly testing will provide maximal opportunity for seroconversion because it provides an adequate amount of time for potentially infected colony animals to shed the pathogens, expose the sentinels, and allow for the sentinel animals to develop a detectable immune response. Skin scraping and faecal floatation test or cellophane tape test will be performed to detect ectoparasites and endoparasites. Generally, one sentinel per cage per rack or side of rack will be euthanized and bled for serology. The serum will be banked for retesting if needed. When sentinel rodents are removed for routine rodent health monitoring, newly acquired sentinel rodents should be replaced within one week. Sentinel cages need to be placed with new sentinel animals every three months.

A summary of the Rodent Health Monitoring results will be sent via email to all rodent users. In the event of positive results, the stored serum sample will be retested for any unexpected positive or equivocal serology. Once outbreak is confirmed, quarantine of the effected colony may be needed and steps may be taken to eliminate/control the agent. A "zone of suspicion" may be established around the affected area. Suspect rooms will be under increased scrutiny. Decisions on how to handle the outbreak will be made in consultation with rodent users. It may be necessary to rederive or depopulate and restock the colony.

A sentinel programme has been implemented at DES for rodent health monitoring of breeding and experimental rodent colonies. Outbred, eight weeks-old female rodents are used as sentinel animals. Sentinel mice used include SPF Swiss Albino from NUS, Swiss Webster mice from Charles River Laboratory and SPF Sprague-Dawley rats from NUS.

Sentinel mice or rats are housed on dirty, soiled bedding from cages of experimental rodents to be sampled. These sentinels are housed in filter-topped cages like all other rodents in the room. One sentinel cage with three sentinel mice are assigned per each side of the rack for mice housed in conventional system and one cage per rack for SPF mice. For rats, a cage of two sentinel rats is placed per rack. All sentinel cages are placed on the lowest shelf, at the end nearest the room's exhaust duct. They are housed in the colony for at least 3 months before being tested.

Cages containing sentinel rodents are specifically identified by label as "SENTINEL PROGRAMME" cage with colour-coded cage card. All sentinels are identified by using ear tags. Sentinel rodents (at least eleven sentinel mice and three sentinel rats) are sent alive to Murine Virus Monitoring Service, 101 Blacks Road, Gilles Plains 5086, South Australia for comprehensive serologic evaluation and extended tests. This is done once a year. At other times of the year (quarterly), sentinel rodents are humanely euthanized and serum collected for basic serologic evaluation. The pelage and caecal contents are also examined for ectoparasites (eg fur mites) and endoparasites (eg pinworms).

All sentinels should be shown to be free of (that is, to test negative for) the agents listed below. The pathogens to be tested quarterly and yearly are described separately for mice and rats as follow:

1. *Basic serology tests for mice (Quarterly)*

- MHV — Mouse Hepatitis Virus
- PARV — Parvovirus (rNS1 antigen)
- ROTA — Rotavirus
- MCMV — Murine Cytomegalovirus
- TMEV — Theiler's Murine Encephalomyelitis Virus
- PVM — Pneumonia Virus of Mice

2. *Comprehensive tests for mice (Yearly)*

- MHV — Mouse Hepatitis Virus
- PARV — Parvovirus (rNS1antigen)
- ROTA — Rotavirus
- MCMV — Murine Cytomegalovirus
- TMEV — Theiler's Murine Encephalomyelitis Virus
- PVM — Pneumonia Virus of Mice
- SEND — Sendai Virus
- PULM — Pneumonia Virus of Mice
- MAD — Mouse Adenovirus type 1
- REO — Reovirus type 3
- LCM — Lymphocytic Choriomenigitis Virus
- ECT — Ectomelia Virus (Mousepox)
- HAN — Hantan Virus
- POLY — Polyoma
- ECUN — *Encephalitozoon cuniculi*

- CAR — Cilia Associated Respiratory Bacillus
- TYZ — *Clostridium piliformis* (Tyzerr's)

3. *Basic serology tests for rats (Quarterly)*

- RCV — Rat Corona Virus
- PARV — Parvovirus
- TMEV — Theiler's Murine Encephalomyelitis Virus
- PVM — Pneumonia Virus of Mice
- TYZ — *Clostridium piliformis* (Tyzer's)
- PULM — *Mycoplasma pulmonis*

4. *Comprehensive tests for rats (Yearly)*

- RCV — Rat Corona Virus
- PARV — Parvovirus
- TMEV — Theiler's Murine Encephalomyelitis Virus
- PVM — Pneumonia Virus of Mice
- TYZ — *Clostridium piliformis* (Tyzer's)
- PULM — *Mycoplasma pulmonis*
- REO — Reovirus type 3
- SEND — Sendai Virus
- LCM — Lymphocytic Choriomenigis Virus
- HAN — Hantan Virus
- ECUN — *Encephalitozoon cuniculi*
- CAR — Cilia Associated Respiratory Bacillus

Extended tests performed include clinical examination, dissection, description of gross lesions and report, histopathology (H & E stain) on lung, liver, kidney, small intestine and colon, culture for *Salmonella*, other enteric pathogens and respiratory pathogens excluding *Mycoplasma spp,* ectoparasite examination, skin scrape for detection of burrowing mites, cellophane tape test for pinworm ova, direct examination of intestinal tract for helminths, wet mount examination of mucosal scrapings for protozoa.

If evidence of murine viruses, parasites or bacterial pathogens is discovered and confirmed in the sentinel animals, investigators are notified for further action. For example, if the sample is positive for pinworm, the researchers will be notified of the result and the entire rodent colony will be given Fendendazole-treated rodent pellets.

CHAPTER 4

BASIC ANIMAL INVESTIGATIVE METHODS

CHAPTER 4.1

BIOIMAGING IN ANIMALS

David Ng, Sidney Yu, S. Somanesan, Manjing Lin, Lin Zheng, Lai Chun Ong, Irene Kee and Choon Hua Thng

PART I: MicroPET

Positron Emission Tomography (PET) is a diagnostic imaging method, which uses a PET camera to measure the concentration and movement of a radiotracer in the living body. A PET scan is performed by a PET scanner that shares many engineering characteristics with other imaging modalities such as computed tomography (CT) and magnetic resonance imaging (MRI). Like CT and MRI, PET scanner reconstructs an image from acquired projections using computers. In PET, the images of the internal structure and functions of the body represent the distribution of a radiopharmaceutical (tracer) within the organs of interest.

Positron-emitting nuclides are neutron-deficient isotopes that achieve stability by the emission of a positive electron, or positron. During the radioactive decay process within the nucleus, the nuclear transmutation of a proton into a neutron results in the emission of a positron and a neutrino. The positron is the antiparticle or the antimatter counterpart of the electron. The positron has an electric charge of +1 and the same mass as an electron. Positrons at the end of their range have the physical property of annihilating with negative electrons to produce a back-to-back emission of two gamma rays. When a positron annihilates with an electron, their rest mass is converted into energy in the form of two gamma ray photons. These two gamma rays travel with energy of 511 keV in opposite directions along the same axis. Twin gamma ray detectors, coupled electronically, provide a signal only when simultaneously triggered by two gamma rays on the same axis. The PET scanner was designed to take advantage of the simultaneous emission of the two annihilation photons. These features provide PET with unique advantages for localizing the source of emission of radiotracer in tissue and organs of the body.

A further advantage of this imaging technique is the very short physical half-life of several positron emitting elements, which ensures low radiation exposure during imaging. Moreover these positron emitting radionuclides such as ^{15}O, ^{13}N, ^{11}C and ^{18}F act as metabolic tracers, which facilitate a wide range of studies of biochemical processes in living tissues and organs. These radionuclides can be incorporated into a wide variety of compounds ranging from small molecules such as water and glucose to complex biochemicals such as peptides, drugs or proteins. ^{18}F, with its longer half life of 110 minutes, allows complex synthesis and fluorine is similar to the hydroxyl group found in almost every biomolecule in size, high C-F bond energy and high electronegativity. Hence, ^{18}F can be used to replace hydroxyl groups without changing its *in vivo* behaviour drastically. So far ^{18}F FDG is the most successful PET radiopharmaceutical. It has a wide range of application in oncology, neurology, and cardiology. Other PET radiopharmaceuticals under development are listed in Table 4.1.1.

The Department of Experimental Surgery at Singapore General Hospital has recently installed a high resolution dedicated PET scanner (microPET™ from Siemens Medical) suited for studying small animal models of human diseases. The department welcomes researchers to use this state-of-the-art research tool in their research. The department would ensure delivery of radiotracer and provide necessary expertise and support for any collaborative research involving microPET. The Department of Nuclear Medicine & PET at Singapore General Hospital operates a cyclotron and a GMP certified PET radiopharmacy and Quality Control laboratories. In addition to the routinely used F-18 based FDG the department has plans to develop and characterize novel molecular imaging probes in the near future.

Basics of PET imaging

The microPET is based on detector blocks consisting of fiber-optic readout of Lutetium Oxyorthosilicate (LSO) scintillator elements using compact multi-channel (that is position-sensitive) photomultiplier tubes (PMTs). It consists of four rings of such position-sensitive scintillation detectors, with 24 detector blocks per ring and therefore a total of 96 detector blocks. Each detector block is cut and acid-etched into an 8 by 8 array of small LSO crystals ($2 \times 2 \times 10$ mm) coupled via optical fibers to a multi-channel PMT. The detector ring diameter of the Rodent microPET is 14.8 cm and its animal port is approximately 12 cm in diameter with an imaging field-of-view of approximately 10 cm transaxially and 8 cm (4-ring) axially. The microPET has no septa and therefore operates exclusively in a 3-D imaging mode. There is a computer-controlled animal bed, a removable point source holder for attenuation correction, a laser alignment system, a Pentium-based computer, electronics and software for data acquisition, data correction, image reconstruction, image display and simple image analysis. The linear and volume spatial resolution of the microPET is approximately 2 mm and $2 \times 2 \times 2$ mm = 8 mm^3 respectively. This is an improvement over existing clinical PET scanners of a factor of approximately 2.2 and 10, respectively. The dimension of the gantry is 76.2 cm (height) × 76.2 cm (width) × 12.7 cm (depth) and it sits on an electronics cabinet measuring approximately 76.2 cm (height) × 106.7 cm (width) × 66.0 cm (depth). The specification of the microPET scanner is mentioned in Table 4.1.2. A picture of microPET scanner and a sample image acquired at the Department of Experimental Surgery are in Figs 4.1.3 and 4.1.4 respectively.

The small animal is positioned within the microPET scanner and the annihilation gamma rays from positron annihilation are captured in coincidence by opposing detectors. The pairs of coincident photons, or events, that are detected, are stored in matrices, or sinograms, where each row in the matrix represents a projection of the activity distribution in the animal. An image reconstruction algorithm is applied to the raw sinograms to process the underlying radioactivity distribution, thus indirectly mapping the functional processes that created the distribution. When the radiopharmaceutical is ^{18}F-FDG, images of FDG accumulation throughout the body are related to tissue glucose utilisation. The basis of PET is therefore the pharmaceutical interacting with the body through a metabolic process; the radioactive-tag allows that interaction to be followed, mapped and quantitated.

Radiotracers for Positron Emission Tomography (PET)

As a unique molecular imaging technique, PET has overcome many of the shortcomings associated with the competing modalities. Modern PET machines provide images with outstanding spatial resolution and can examine the entire body with an exquisite detail in a reasonably short period of time. Among functional imaging techniques, PET stands out as a powerful modality that can generate the most accurate quantitative results; as such it can play a critical role in research and clinical applications.

PET technology is a multidisciplinary science. It involves chemistry, physics, biology, and medicine. Over the past three decades, developments in radiotracer chemistry and positron emission tomography instrumentation have merged to make positron emission tomography a powerful scientific tool in the biomedical sciences. Radiotracer chemistry has played a pivotal role in driving the field in new directions in studies of human physiology. At the heart of the development of radiotracer chemistry is synthetic chemistry, which directed to the rapid incorporation of simple short-lived precursor molecules into organic compounds that can be used to map specific biochemical processes and the movement of drugs in living systems.

1. *Radiotracer chemistry*
   Time dominates all aspects of a PET study. PET tracers must be synthesized and imaged within a time frame compatible with the half-life of the isotope. Large amount of radioactivity need to be used to compensate for radioactive decay and for the sometimes-low synthetic yields. Thus, shielding, remote operations, and automation are integrated into the experimental design, and automated or semiautomated systems or black boxes are developed for radiotracers in routine use. It is ideal to introduce the radioactivity at the last step in the synthesis, which may require a multi-step synthesis of a suitably protected precursor compound into which the radioactivity isotope can be introduced. Protective group must be stable to the labeling conditions and readily removed. The crude reaction mixture is usually purified using a high-performance liquid chromatography or a combination of solid-phase extraction and high-performance liquid chromatography. Because PET radiotracers are typically administered intravenously, procedures must be developed to yield pharmaceutical quality materials.

Currently, DES is developing a radiotracer chemistry laboratory. One of the major current and future concerns of the department is to attract and train chemists in this field, in which advances in chemistry are often the rate-limiting step in advancing new knowledge in biology and medicine for PET research. Therefore, the department would ensure delivery of various radiotracers on demand to support the research for microPET, which has recently been installed in the department.

2. *Specificity of ^{18}F-fluorodeoxyglucose (^{18}F-FDG)*
Among the developed radiotracers, ^{18}F-fluorodeoxyglucose (^{18}F-FDG) has been the most widely used radiotracer for PET studies in neuroscience, cardiology and oncology. The ^{18}F-FDG technique was introduced in 1976, and the effectiveness of this modality as a molecular probe has been effectively demonstrated in the investigation of a multitude of serious disorders. This agent was proposed as a novel tracer to determine regional brain function in normal physiologic states and in neuropsychiatric disorders. With the introduction of instruments that are capable of rapid whole-body imaging, this critical role of PET imaging with ^{18}F-FDG in the management of a number of serious disorders has been well established. So far, However, tissue uptake of ^{18}F-FDG is nonspecific and this uptake can increase due to inflammation. Thus, more specific radiotracers are desirable.

3. *Radiotracers beyond FDG*
Various positron-emitting radiotracers have been developed for PET studies. However, the potential for synthesizing numerous biologically important compounds with positron-emitting radionuclides, such as carbon-11 (^{11}C) and fluorine-18 (^{18}F), has yet to be fully realized. In addition, positron-emitting metallic radionuclides, such as Technetium-94 (^{94m}Tc) and Copper-64 (^{64}Cu) can be used for diagnostic purposes and may further expand the domain of PET for functional studies.
The great majority of the PET tracers are labeled with carbon-11 (^{11}C) and fluorine-18 (^{18}F). Over the past two decades, a variety of PET radiotracers have been developed with ^{11}C and ^{18}F for application in neuroscience and oncology. The radiotracers for application in neuroscience include those used for studying dopamine, serotonin, opiod, norepinephrine, and the cholinergic system. Recently, radiotracers for imaging β-amyloid also have been developed. The current development of radiotracers for application in oncology target amino acid, protein synthesis, DNA synthesis, cell proliferation, lipid synthesis, receptor-mediated radioligands, hypoxia, angiogenesis, apoptosis and gene therapy. Structures of some of the older and most familiar radiotracers are presented in Fig 4.1.1 and those that have been developed more recently are presented in Fig 4.1.2.

4. *Outlook*
With advances in the sequencing of the human genome, we can expect to identify many new genes and their protein products. This creates a sense of urgency in developing radiotracers for characterizing the functional activity of these new proteins in living systems, ultimately in animals and humans. In addition, new knowledge on progenitor cells and their promise in treating human disease calls for expanding radiotracer development so that imaging can be used to study cell

trafficking as well as the molecular process involved in stimulating these cells to differentiate in vivo. From the late 1990s, the development of radiotracers to monitor the efficacy of gene therapy and organ transplantation has become a new area of research, one in which work will continue in the foreseeable future.

5. *Radiotracers for PET imaging*

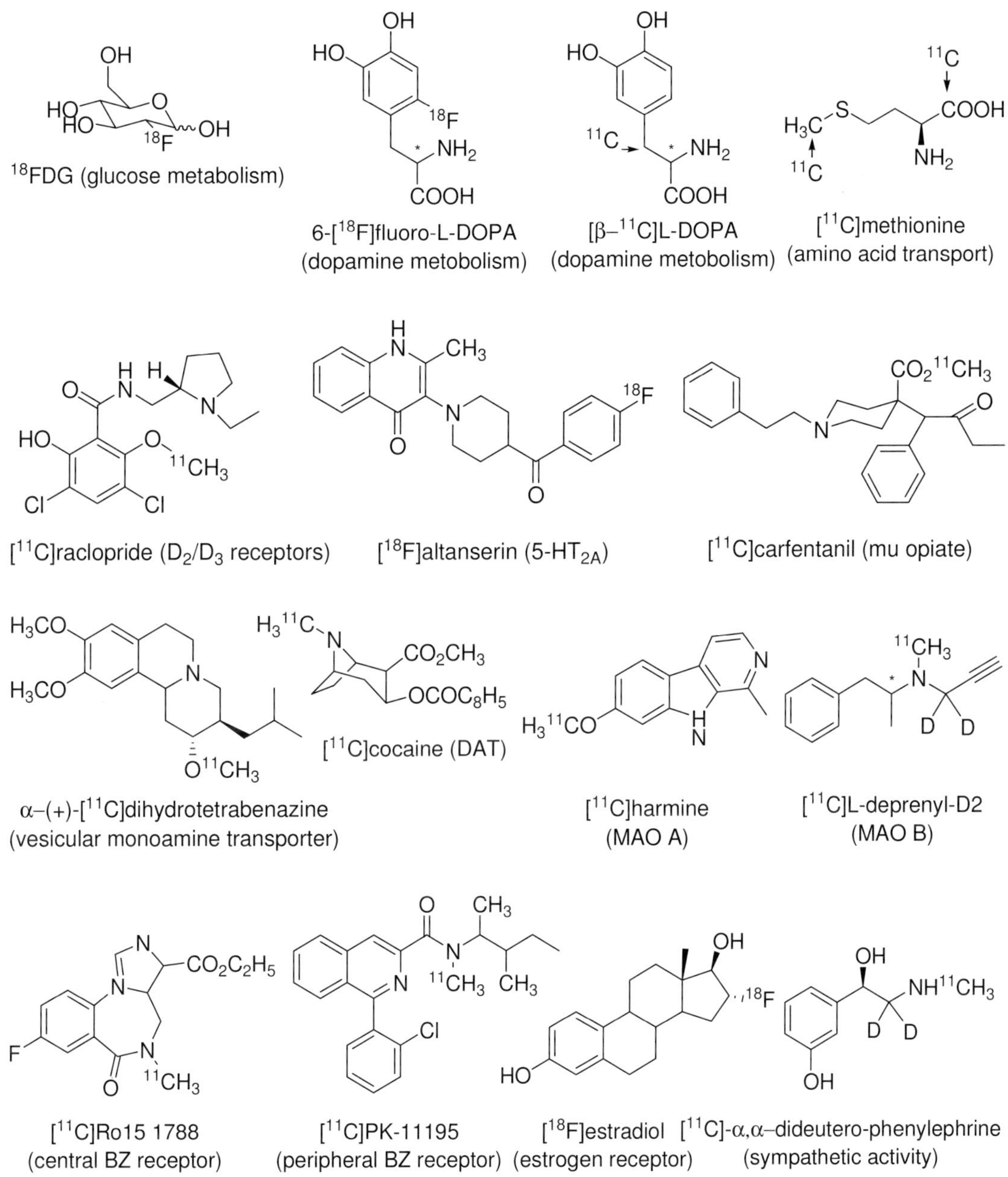

Fig 4.1.1: Structures and molecular targets for some of the older radiotracers that have been under investigation for a number of years.

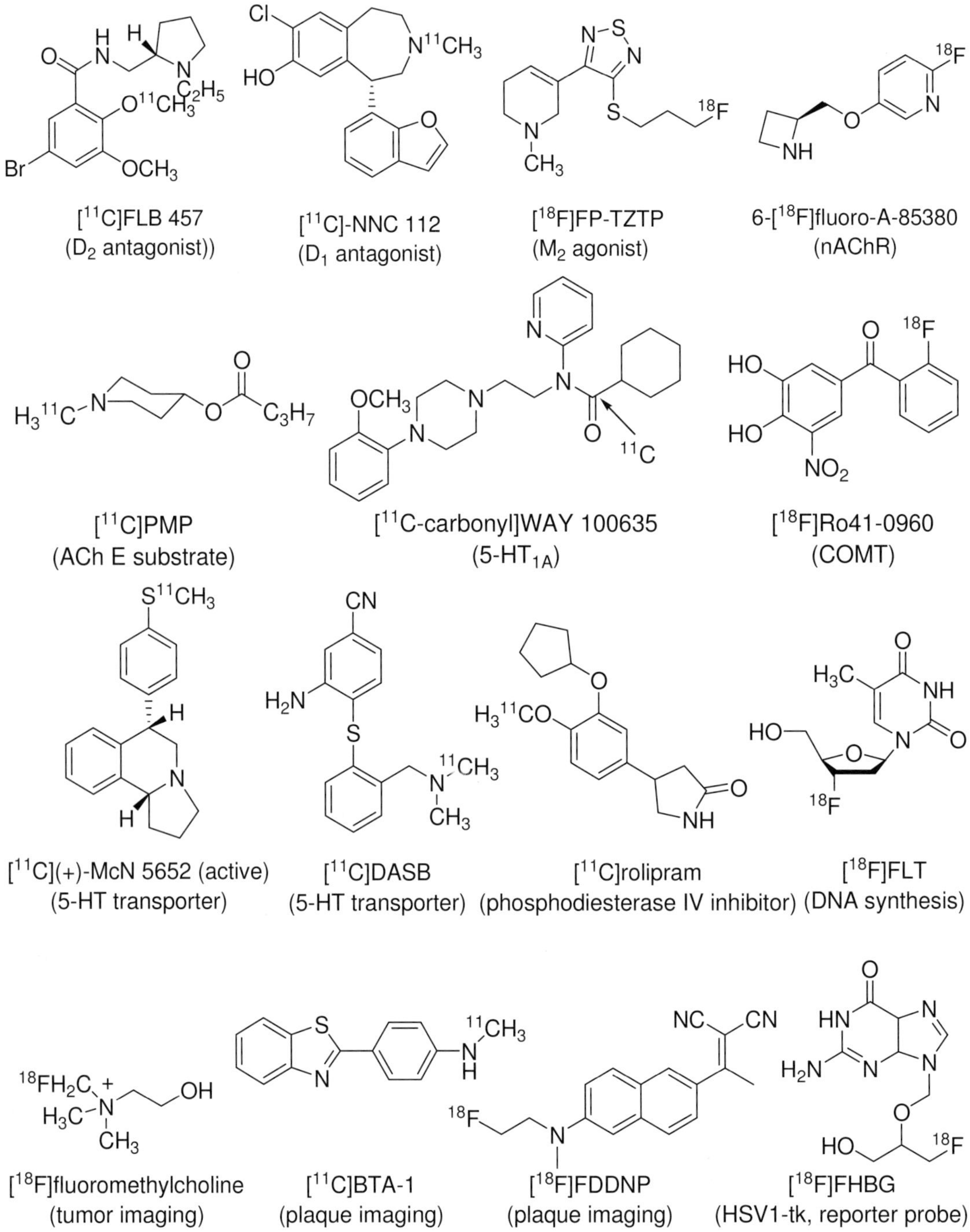

Fig 4.1.2: Structures and molecular targets for some of the newer radiotracers.

Challenges in Small Animal Imaging

Compared to human PET scanning, microPET presents new challenges, both of instrumentation and biological nature.

A. *Instrumentation related challenges*

One of the greatest instrumentation challenges that had to be overcome was to increase spatial resolution while still maintaining high detection sensitivity. This is because the size of a mouse organ is orders of magnitude smaller compared to the size of the corresponding human organ. As such, the spatial resolution of a traditional human PET scanner would not be sufficient for microPET.

In PET, radiotracer decay is measured by detecting in temporal coincidence the two 511 keV gamma rays originating from positron decays. Thus small detectors with high gamma stopping efficiency were needed to achieve the desired performance outcomes. The introduction of Lutetium Orthosilicate (LSO) as the gamma detection material have made it possible for the size of individual crystals to be reduced to dimensions of the order of a couple of millimeters due to its high density and light yield. This leads to spatial resolutions in the range of 1 mm^3 to 1.8 mm^3 while still maintaining high detection sensitivity.

B. *Biology related challenges*

At least two other biological related features exist in animal scanning: the small physical size of the animals and the need to administer anaesthetics. The small sizes of the animals place a limitation to the amount of tracer that can be administered because the administered radiotracer must not influence the process under investigation. This requires tracers to be highly specific and also restricts the amount of radioactivity that can be injected. The latter thus renders detection sensitivity even more critical.

Similar to a human subject in a clinical scanner, the animal must remain still and the higher the resolution of the system, the smaller the movement that can be tolerated. Therefore, anaesthesia remains a prerequisite of scanning. It is therefore inevitable that the physiology of the animal will be abnormal due to action of the anaesthetics effects during the scan. Other challenges that have direct implications on PET imaging include data interpretation of small animals and quantitative analysis.

Applications of Molecular Imaging with MicroPET

The last several decades have seen the vast growth in the fundamental understanding of biological and disease phenomena in terms of molecular events. With the development of molecular medicine, the level of complexity has shifted from macroscopic disease entities to elucidating the cellular and subcellular mechanisms of disease. Along with this evolution, new paradigms of diagnosis, therapeutics, management have emerged. Imaging at the level of cellular and subcellular processes has become more sophisticated and potentially critical in detection and intervention of disease at an early stage. Among the methodologies that image at the cellular/molecular level are bioluminescence imaging, optical imaging, MRI, PET and single-photon emission tomography. There are more than 250 centers worldwide using PET scanners in clinical studies, and there are at least 8 operational MicroPET systems. MicroPET is a noninvasive system that eliminates the need for biopsies and thereby extends an animal's life. It allows serial and longitudinal studies to be performed on the same living animal, enabling researchers to follow a single animal over time and monitor the effects of interventions on disease progression and outcome. MicroPET will be particularly valuable for studying genetically modified animals that exhibit high variability or are unique or valuable.

MicroPET offers the unique opportunity to image small animal models of diseases, including genetically engineered animals. It is a functional imaging modality at molecular level and provides valuable insights into biochemical, physiological, pathological or pharmacological process *in vivo*. Data can be obtained noninvasively, repeatedly, and quantitatively in the same animal. Current applications include a diverse field including perfusion, metabolism and substrate utilization in various vital organs including heart and brain, gene expression and stem cell tracking, neurotransmitter and receptors, neural activation and plasticity, targeting tumour antigens and elucidating tumour biology such as angiogenesis, hypoxia and apoptosis. Recent research efforts find its application in a wide area ranging from basic insights into the normal physiology and disease processes to drug development and early response to anticancer and gene therapy.

In order to demonstrate some of the efficacy of microPET imaging in small-animal studies and to demonstrate how microPET has been used in animal research, the following discussion will focus on a few illustrative case examples of the use of microPET imaging of small-animals.

Use of Animal Models as Disease Investigative Tool – Possibility of Interventions

MicroPET has been used quantitatively to investigate functional changes occurring as a consequence of a disease or specific intervention. Since longitudinal studies on the same animals are now possible with microPET, it becomes feasible to perform more sophisticated studies, such as investigating chronic vs. acute effects of treatment or simply effects of disease as approximated by the particular animal model. In particular, numerous studies have shown the potential of small animal PET imaging for the analysis of animal models of central nervous system disorders such as Parkinson's disease. By using a suitable radiotracer, microPET is able to quantify the differences in the striatal lesioning between a healthy, a moderately and a severely lesioned mouse. In addition, it is also now becoming common to use microPET imaging in tumour models in mice to investigate tumour proliferation, progression and treatment.

Testing of New Drugs and Their Efficacy

MicroPET is an ideal tool in the process of new drug development and evaluation of treatment efficacy. By directly radiolabeling the novel drug, microPET imaging allows the biodistribution to be dynamically monitored in single animals. The location, concentration, and time course of the drug can then be evaluated. Another approach is possible for drugs designed for targets for which PET-labeled probes already exist. PET imaging is then performed to measure the ability of the drug to compete with the PET probe for the target, thus allowing a measure of the *in vivo* delivery and affinity of the drug. Finally biomarkers such as blood flow, glucose metabolism or cell proliferation can be used to measure pharmacodynamics effects. This is especially useful when PET tracers are already available for the process that the drug is supposed to modify. One such example would be the use of blood flow tracers to measure angiogenic response.

1. *Case example 1*
   Non-invasive imaging of islet grafts using positron-emission tomography
   Lu Y, Dang H, Middleton B, Zhang Z, Washburn L, Stout D B, Campbell-Thomson M, Atkinson MA, Phelps M, Gambhir SS, Tian J, Kaufman DL.
   Proc Natl Acad Sci USA 2006; 103: 11294-11299.

   Islet transplantation offers a potential therapy to restore glucose homeostasis in type 1 diabetes mellitus. Islet transplantation is not routinely successful because most islet recipients gradually lose graft function. Serological markers of islet function are insensitive to islet loss until late stages. Islets have been engineered to express firefly luciferase using bioluminescence imaging. However, this is currently limited to tissues not deeper than 2 cm.

   PET imaging was used by engineering an adenovirus carrying a reporter gene into islet tissues and non-invasive repetitive imaging of this reporter gene with microPET. NOD/SCID mice were transfected with recombinant adenovirus bearing the mutant reporter gene HSV1-sr39tk and was imaged with positron-labelled [F-18] 9-[(4-fluoro-3-9 hydroxymethyl) butyl] guanine ([F-18] FHBG). Findings showed that the PET signals from the reporter genes was directly related to the implanted islet mass.

2. *Case example 2*
   ***In vivo* imaging of neuronal activation and plasticity in the rat brain by high resolution positron emission tomography (MicroPET)**
   Kornblum HI, Araujo DM, Annala AJ, Tatsukawa KJ, Phelps ME, Cherry SR.
   Nat Biotechnol 2000; 18: 655-660.

   MicroPET was used for the mapping of stimulation responses in the brain. This study demonstrated that microPET can be used to assess metabolic activity in different regions of the conscious rodent brain using [18F] fluorodeoxyglucose (FDG) as the tracer, and to monitor changes in neuronal activity.

   Limbic seizures result in dramatically elevated metabolic activity in the hippocampus, whereas vibrissal stimulation results in more modest increases in FDG uptake in the contralateral neocortex.

   MicroPET can be used to study lesion-induced plasticity of the brain. Cerebral hemidecortication resulted in diminished relative glucose metabolism in the neostriatum and thalamus ipsilateral to the lesion, with subsequent, significant recovery of metabolic function.

   These studies demonstrate that microPET can be used for serial assessment of metabolic function of individual, awake rats with a minimal degree of invasiveness, and therefore, has the potential for use in the study of brain disorders and repair.

3. *Case example 3*

   Longitudinally quantitative 2-deoxy-2-[(18)F]fluoro-D-glucose micro positron emission tomography imaging for efficacy of new anticancer drugs: A case study with bortezomib in prostate cancer murine model

   Zhang Y, Saylor M, Wen S, Silva MD, Rolfe M, Bolen J, Muir C, Reimer C, Chandra S.

   Mol Imaging Biol 2006; 8(5): 300-308

The aim of this study was to validate quantitative metabolic response of tumors to a treatment measured by longitudinal 2-deoxy-2-[(18)F] fluoro-D-glucose (FDG) micro positron emission tomography (microPET) as a robust tool for preclinical evaluation of new anticancer agents.

Severe combined immunodeficiency mice with CWR22 xenografts were intravenously treated with bortezomib (Velcade) at 0.8 mg/kg on days 0, 3, 7, 10, and 14 and imaged with FDG microPET before, during and after treatment. Quantitative indices of tumor FDG uptake were developed.

FDG microPET images successfully revealed the gradual reduction of tumor FDG uptake on day 4 onward despite no absolute tumor shrinkage. The standardized uptake values of FDG in tumors were reduced to 43 % of the baseline values. Using the total tumor FDG uptake as the viable tumor burden, we found 86 % tumor inhibition, compared to a 55 % tumor growth inhibition in tumor volume measurement. FDG microPET imaging can provide an additional dimension of the efficacy of anticancer therapies that may otherwise be underestimated by tumor volume measurement.

4. *Case example 4*

   Monitoring of therapy in androgen-dependent prostate tumor model by measuring tumor proliferation

   Oyama N, Ponde DE, Dence C, Kim J, Tai Y-C, Welch MJ.

   J Nucl Med 2004; 45: 519–525.

3-Deoxy-3-18F-fluorothymidine (18F-FLT) has been recently described as a radiopharmaceutical for measuring cellular proliferation using PET imaging. Evaluation of tumor proliferative activity by PET using 18F-FLT could be a procedure to assess the viability of tumor, such as histologic grade, clinical stage, and prognosis as well as the early effects of cancer therapy.

This study was undertaken to determine whether 18F-FLT is useful in the detection of prostate cancer as well as monitoring therapeutic effects in a human tumor model. The androgen-dependent human prostate tumor, CWR22, was implanted into athymic mice. Androgen ablation studies were conducted in the CWR22 model with either diethylstilbestrol (DES) or surgical castration. The effectiveness of therapy was monitored using 18F-FLT microPET as baseline, during treatment, and after treatment.

MicroPET using 18F-FLT successfully imaged the implanted CWR22 tumor in the mice at both 1 and 2 h after injection. There was a marked reduction of 18F-FLT uptake in tumor after castration or DES treatment; however, there were no differences in 18F-FLT uptake in the tumor in the control group. These changes of 18F-FLT uptake in tumor parallel the changes of actual tumor measurement. These results indicate that microPET with 18F-FLT is useful for detection of prostate cancer in an animal model. 18F-FLT has the potential for monitoring the therapeutic effect of androgen ablation therapy in prostate cancer.

5. *Case example 5*
Evolution of diaschisis in a focal stroke model
Carmichael ST, Tatsukawa K, Katsman D, Tsuyuguchi N, Kornblum HI.
Stroke 2004; 35: 758-763.

Stroke produces diaschisis in adjacent and connected regions. The sequential changes in diaschisis over time and the relationship of regions of diaschisis to functional cortical areas and regions of poststroke neuroplasticity have not been determined.

Small cortical strokes were produced in the barrel cortex of rats. Relative glucose metabolism was determined *in vivo* over time with [18F] fluorodeoxyglucose small-animal positron emission tomography. Cerebral blood flow was measured with [14C] iodoantipyrine. Regions of hypometabolism and hypoperfusion were compared with histological damage in the same animals.

Small cortical strokes produce an initial network of hypometabolism in a broad region of cortex adjacent to the stroke and in the striatum and thalamus on day 1. Cerebral blood flow is diminished only immediately around the cortical infarct on day 1. A substantial area of cortex adjacent to the stroke remains hypometabolic on day 8. This persistent cortical hypometabolism occupies the somatosensory cortex, forelimb motor cortex, and second somatosensory area. Focal stroke produces ipsilateral diaschisis in connected cortical regions that is clearly distant from subtotal damage and may play a role in poststroke neuroplasticity.

6. *Case example 6*
^{18}F-FDG Small-Animal PET for Monitoring the Therapeutic Effect of CT-Guided Radiofrequency Ablation on Implanted VX2 Lung Tumors in Rabbits
Okuma T, Matsuoka T, Okamura T, Wada Y, Yamamoto A, Oyama Y.
J Nucl Med 2006; 47: 1351-1358

The primary goals of this study were to investigate the behavior of normal lung tissues after radiofrequency ablation (RFA) and to determine the suitability of ^{18}F-FDG PET, using a dedicated small-animal scanner, for monitoring the early therapeutic effects of RFA on VX2 lung tumors (VX2s) in rabbits.

Fourteen Japanese white rabbits with normal lungs underwent RFA, followed by ^{18}F-FDG PET at 1 d and at 1, 2, 4, and 8 wk. In addition, 7 rabbits with untreated VX2s underwent ^{18}F-FDG PET, and 13 rabbits with RFA-treated VX2s underwent ^{18}F-FDG PET at 1 d ($n = 7$) or 1 wk ($n = 6$) after the treatment.

For ablated tumors, values of uptake were significantly lower than those for nonablated tumors ($P < 0.001$). Histopathologic examination confirmed the absence of viable tumors. This study demonstrates that ^{18}F-FDG PET is promising for evaluating the therapeutic response of lung malignancies to RFA.

This is a time of paradigm change in biology, biotechnology and medicine, in which molecular events are the fundamental drivers of biological and pathological phenomena. MicroPET is one of several tools well-suited to explore and investigate these phenomena at the molecular level. Table 4.1.1 shows probable PET radiopharmaceuticals and their potential application. Table 4.1.2 lists the specifications of microPET machine located at Department of Experimental Surgery.

Table 4.1.1: Probable PET radiopharmaceuticals and their potential application

Radiopharmaceutical	Potential application
1. Oncology	
F-18 FCH	Prostate cancer
F-18 FLT	Cell proliferation
F-18 DOPA	Amino acid transport
F-18 FET	Amino acid transport
F-18 PFBG	Neuroendocrine tumour
F-18 Misonidazole, F-18 FAZA, F18-PIMO	Hypoxia
F-18 Fluoro-etanidazole	Hypoxia
F-18 MDHT	Androgen receptors
F-18 FES	Breast cancer
F-18 NaF	Bone lesions
F-18 FACBC	Amino acid transport
F-18 Monoclonal antibody	Tumour imaging
2. Neurology	
F-18 DOPA	Dopaminergic terminals density
F-18 Fallypride	Dopaminergic system D2 receptors
F-18 MPPF	Serotoninergic system 5-HT1a receptors
F18 Fluoroethyl spierone	Dopaminergic system D1 receptor
3. Cardiology	
F18 Fatty acids and fluoropalmitate	Energy substrates metabolism
F-18 PFBG	Detecting neuronal damage
4. Gene Imaging	
F-18 FHBG	Reporter gene imaging
F-18 FHPG	Reporter gene imaging

Table 4.1.2: Specifications of the microPET located at the Department of Experimental Surgery

Detector Diameter	15 cm
Transverse Field of View (Animal Port Diameter)	12 cm
Axial Field of View	8 cm
# of Detector Elements per Detector Block	$8 \times 8 = 64$
Dimensions of Detector Element	1×1 mm
# of Detector Blocks per Detector Module	4
Total # of Block Rings	4
Total # of Detector Blocks per Block Ring	24
Total # of Detector Blocks	$24 \times 4 = 96$
Total # of Detector Elements	$64 \times 24 \times 4 = 6144$
Dead Time (Integration Time per Block)	~300 nsec
Coincidence Resolving Time	3.2 nsec FWHM
Reconstructed Spatial Resolution	2 mm FWHM
Reconstructed Volume Resolution	$2 \times 2 \times 2$ mm = 8 mm^3
Sensitivity	900 cps/μCi

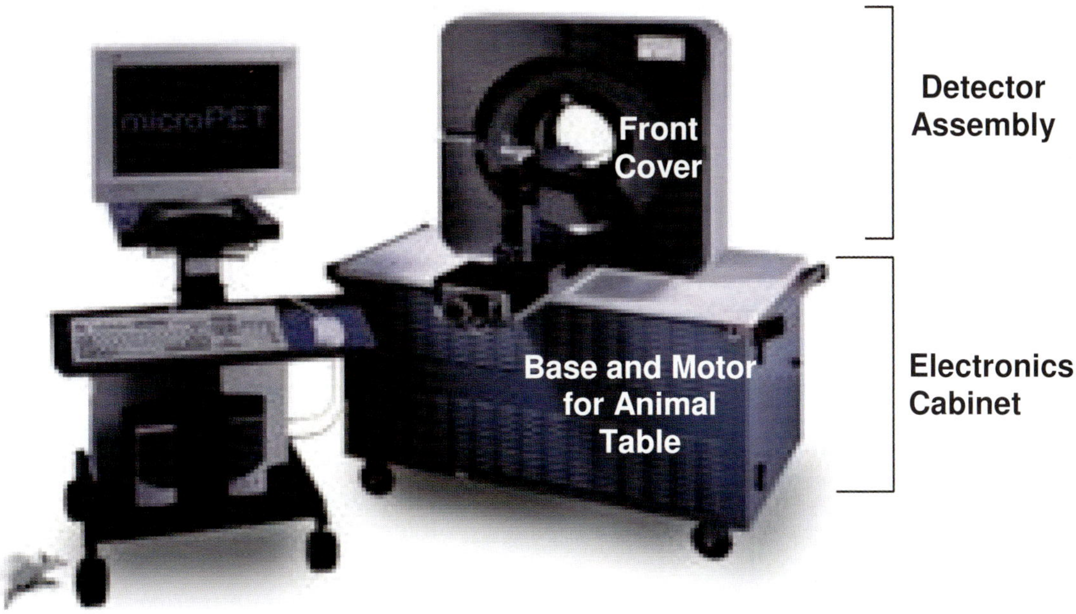

Fig 4.1.3: MicroPET scanner with the animal imaging table and processing computer.

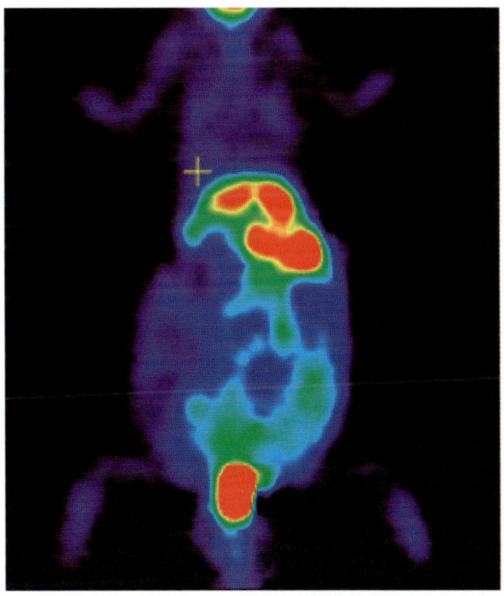

Fig 4.1.4: Sample image of a small animal PET image taken at Department of Experimental Surgery, Singapore.

PART II: MicroCT

The microCT (micro computed tomography) is one of the dedicated high-resolution small animal imaging modalities that have recently emerged as an important new tools for laboratory animal research.

This imaging system permits researchers to noninvasively screen animal models for mutations or pathogenesis and allow for long term observation of a single animal in response to therapy as well as to monitor its disease progression.

Rigaku R_mcT Series microCT (RmCT)

The Rigaku R_mcT series microCT (Fig 4.1.5) is developed and manufactured in Japan by the J Morita Manufacturing Corporation. It utilises a 17 second ultra high speed volume scan with a high resolution (20 to 135 μm) and a FOV of $48 \times 48 \times 36$ mm (max) at 1 MegaVoxels. The basic specifications are as follows:

X-ray generator:	Tube voltage: <90kV Tube current: <200uA
Scanning method:	Rotate-Rotate method with cone beam scan
Voxel size:	Approx.135, 100, 50, 20 um
Measurable size:	About $\varnothing$ 180×240 mm
Imaging volume:	About $\varnothing$ 65×65 mm - $\varnothing$ 10×10
Processing time *1:	Volume scan: About 17 sec Image reconstruction: About 2 min
Overall size:	1400(W) $\times$ 900(D) $\times$ 1600(H) mm
Weight:	570 kg
Power supply:	AC100 V, 5A
X-rays leakage:	Less than 1 uSv/hour
Safety mechanism:	Door interlock Safety key switch Emergency stop switch
Control and data processing:	Windows XP Professional

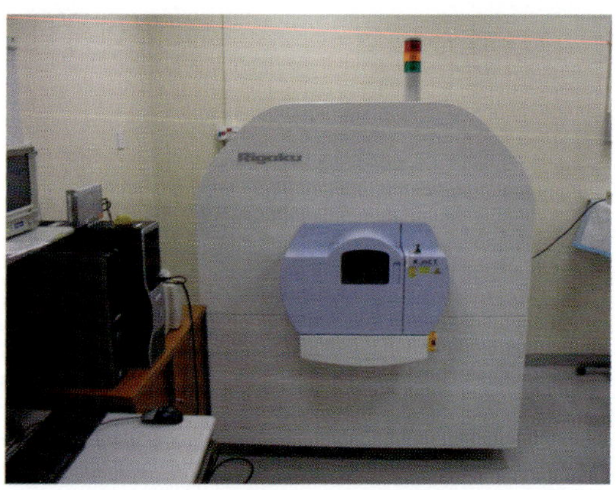

Fig 4.1.5: microCT machine.

Soft Tissue Imaging Using the Fenestra Solution

The microCT has been primarily used for bone imaging applications such as characterizing the 3D trabecular structure and 3D quantification of modelling and remodelling processes.

Researchers and scientists interested in visualising anatomy in living animals using CT imaging technique techniques are often faced with the problems of the relative lack of soft tissue contrast and the long acquisition time. Alerion Biomedical, Inc. has licensed a technology, which comprised of iodinated lipids that provide contrast enhancement and a novel oil-in-water lipid emulsion that selectively localizes the lipids to various locations within the animal. Alerion's Fenestra^{TM} LC provides visualization of the hepatobiliary system by exploiting the endogenous lipid metabolism pathways present in the animal body. The primary benefits of the agent are:

- Prolonged contrast enhanced of the entire hepatobiliary system from a single administration.
- Intracellular localization in hepatocytes correlated to metabolic status of the liver.
- Simultaneous anatomical and functional assessments possible.
- Repeat administration without safety concerns.
- Pharmacokinetic profile compatible with microCT imaging temporal constraints.
- Physicochemical properties compatible with parenteral administration.

Alerion's Fenestra^{TM} VC, a refined version of Fenestra^{TM} LC provides a contrast enhancement of the entire vascular system from a single administration. Because of their comparatively long and stable *in vivo* residence times (up to several hours), the Fenestra^{TM} solution has shown considerable promise for use in microCT imaging procedures.

The following are the procedure for imaging soft tissues in rodents:

1. Animal Models – Rodents

The following steps are required to prepare the animals for imaging:

a. Fasting/GI preparation: MicroCT imaging is best performed in rodents that have been maintained on a non-chow, liquid diet for 24 hours prior to study so as to reduce the imaging artifacts due to minerals that are found in rodent chows.
b. Tail vein preparation: The rodents are kept in a warm environment (approximately 30 °C) to ensure proper vessel dilation prior to injection of Fenestra^{TM} solution. A similar effect may be achieved using an infrared lamp to warm up the tail.
c. Anaesthesia: Rodents were anaesthesized with an intraperitoneal injection of a cocktail of ketamine (50 mg/kg body weight) and diazepam (5 mg/kg body weight), which provides a 45 to 60 minute anaesthesia.

2. Fenestra^{TM} Dosing

Proper dosing of Fenestra^{TM} is important to be able to obtain good images. Following are the dosing steps employed.

a. Fenestra™ LC or VC was administered at a dose of 20 ml per kg body weight.
b. A 1ml disposable syringe fitted with a 30 gauge needle is ideal for injection of Fenestra solution for mice and a bigger syringe fitted with a 27 gauge needle is ideal for larger rodents such as rats via their tail vein.
c. The injection rate for the Fenestra solution is maintained at 0.4 ml per minute over a period of time.
d. Expected residence time for imaging purposes: 0 to 6 hours post-injection, with peak at 3 to 4 hours post-injection.

3. *Image Acquisition*
To be able to produce the best images, certain steps have to be followed. These steps are listed below:

a. Animal orientation: The anaesthetized rodent was placed on the imaging table in the prone position with its head oriented into the gantry.
b. Pre-contrast exam: A pre-contrast scan of the anaesthetized animal is usually done. This pre-contrast scan will be used as a baseline scan.
c. CT scan setup is as listed in Table 4.1.3.

Table 4.1.3: CT scan setup

Magnitude	1.5	2.0	4.0	10
Recommended Tube Voltage (kV)	90	90	90	90
Recommended Tube Current (uA)	88	50	88	88
FOV (Imaging Area) (mm)	Ø64 × H64	Ø48 × H48	Ø24 × H24	Ø9.6 × H9.6
Resolution (Voxel Size) (um^3)	133 × 133 × 133	100 × 100 × 100	50 × 50 × 50	20 × 20 × 20
Sample Backet Size	L(Ø200) or M(Ø150)	L or M	S (Ø50)	S
Sample Bed Size	L or S	L or S	S	S

Figures are contributed courtesy of J Morita Manufacturing Corporation.

The choice between the various protocols depends on the spatial resolution required. A larger voxel size (133 um^3) would enable imaging of a larger area (64 mm) whereas a smaller voxel size (20 um^3) would enable imaging of a smaller area (9.6 mm).

4. *Data Visualization*

a. i-Dixel : Data is routinely imported from the i-Dixel reconstruction programme as raw CT image data or as bitmaps windowed to a contrast setting determined by the operator. Data can be viewed in DICOM with simultaneous display of the axial, coronal and sagittal images.
b. Exporting images: Planar and 3D images can be exported as raw data in DICOM files format or bitmaps images for use in presentation or publication purposes.

Representative CT Image

Fig 4.1.6 shows a representative CT image, without contrast while Fig 4.1.7 shows a CT image with the use of contrast.

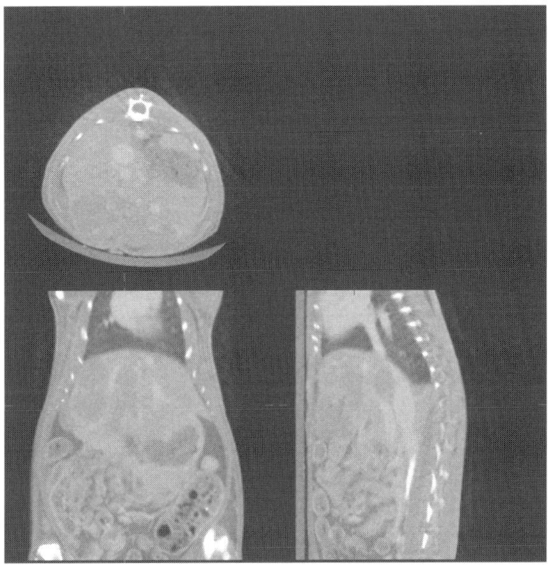

Fig 4.1.6: Noncontrast axial, coronal and sagittal scan of a male rat obtained with RmCT.

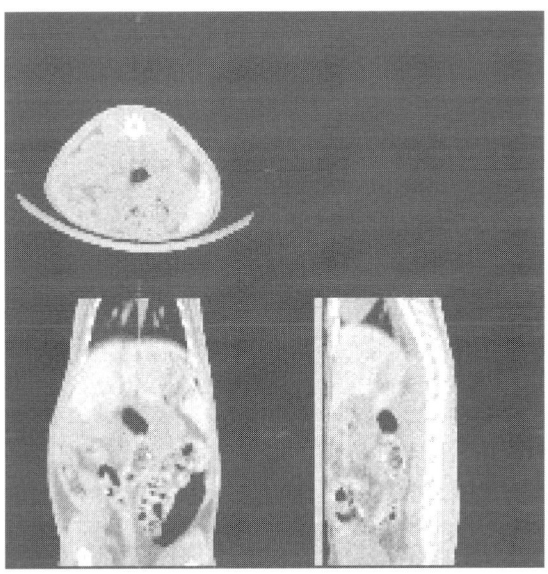

Fig 4.1.7: Axial, coronal and sagittal scan of a male rat 1.5 hr after IV injection of Fenestra™ LC. Note: The liver is enhanced and better tissue contrast is evident in the abdominal cavities.

CHAPTER **4.2**

HISTOLOGY SAMPLING AND TECHNIQUES

In Chin Song

Tissue Fixative

There is a wide variety of tissue preparation and histological techniques available. Fixation is the most important single factor in the preparation of good histologic sections. Prompt and adequate fixation is imperative. Without this, all other histopathological procedures become unnecessarily time consuming and the resulting slides may be non-diagnostic.

A. *Neutral Buffered Formaldehyde*

The most commonly used fixative in histopathology is 4 % formaldehyde, giving 10 % formalin for tissue fixation. 10 % neutral phosphate buffer formalin at pH 7 is also used to harden tissues but produce little or no shrinkage. It is well tolerated by tissues and has good penetration. Formalin fixes not by coagulation but addition, reacting primarily with amino acids to form cross-linking molecules and that the structures of intra-cytoplasmic proteins are not significantly altered. It prevents the process of autolysis and bacterial attack of tissue. Moreover, it makes the tissue more resistant to different stages of chemical treatment during processing. The microanatomy of tissues will be well preserved to allow for accurate diagnosis.

Formula for neutral buffered formaldehyde is provided below:

37 to 40 % Formaldehyde	- 100 ml
Distilled water	- 900 ml
Sodium dihydrogen phosphate monohydrate	- 4 g
Disodium hydrogen phosphate anhydrous	- 6.5 g

Regardless of the fixative used, adequacy of fixation is markedly dependent upon the thickness of the block and the time allowed in fixative. Except under unavoidable circumstances, tissues should not exceed three to four mm in thickness. For 24 hours fixation, 10 to 20 times the tissue volume of neutral buffered formaldehyde is required. Specimens from encapsulated organs such as lymph nodes or spleen should be cut and fixed as soon as possible. Any tissue left exposed to air for a prolonged period, will invariably show a peripheral zone of air-dried cells and central autolysis. One can easily recognize sections from such tissue since the staining varies from the edge to center. The periphery is deeply stained and the central portion is pale. The cells at periphery may be readable, but those centrally are ballooned, distorted, and poorly stained. Prolonged fixation is also known to cause shrinkage and hardening of tissue. It severely inhibits enzyme activity and immunological reactions. For bulky tissues, seven to 10 days is required to achieve better fixation. Tissue is then transferred into 70 % alcohol.

B. *Glutaraldehyde Fixative*

Glutaraldehyde fixative is the most commonly used fixative for Transmission Electron Microscopy (TEM). This fixative provides excellent cytologic fixation. It stabilizes proteins via a cross linking mechanism involving amino groups of lysine and other amino acids through the formation of pyridine intermediaries.

From stock solution:

- 25 % glutaraldehyde stock solution
- Cocodylate buffer 0.1 mol/L, pH 7.4

Sodium cacodylate trihydrate, $Na(CH_3)_2AsO_2.3H_2O$	- 21.4 g
Distilled water	- 1.0 l

The glutaraldehyde and cocodylate buffer are combined in proportions as indicated below:

25 % glutaraldehyde stock solution	- 10 ml
0.1 mol/L cocodylate buffer, pH 7.4	- 90 ml

C. *Fixatives Used for Different Tissues*

Table 4.2.1 lists the fixatives commonly used for different tissues.

Table 4.2.1: List of fixatives used for different tissue specimens

Tissue	Fixative	Time	Explanation/Description
Eye	10 % Formalin	48 hrs	Perfusion technique required to give better fixation
Brain	10 % Formalin	2 wks	Perfusion technique achieves fixation in 5-6 day (Adickers et al. 1997)
Kidney	10 % Formalin	24 hrs	
Gastrointestinal tract	10 % Formalin	24 hrs	
Liver	10 % Formalin	24 hrs	
Lung	10 % Formalin	24 hrs	Inflation with fixative through main bronchi for better fixation
Lymphoid tissue	10 % Formalin	24 hrs	Node required to be split into half for better penetration of fixative
Muscle	10 % Formalin	24 hrs	Allow muscle to relax after harvesting for 10 min before fixation

Treatment of Mineral Hard Tissues

In soft tissues, good fixation usually allows sufficient penetration of antibody complexes into the tissue sections. However, in bone and cartilage, very dense extracellular mineralized matrix can prevent sufficient penetration even after adequate fixation. Bone specimens must be completely fixed in neutral buffered 10 % formalin before decalcification and processing. This helps in protecting the bone and surrounding soft tissue from damage during decalcification. Acid and chelating agents are used for decalcification of bone specimens before chemical processing and paraffin embedding. These include:

A. *Strong Acid*

Strong acid, which decalcifies bone rapidly, contains hydrochloric or nitric acid in low pH. In 24 to 48 hours of decalcification, it causes tissues swelling and damages the tissue stainability especially in uptake of haematoxylin and eosin stain by nuclear chromatin and cell nuclei. It interferes with either substrates or enzymes and invalidates or produces unsatisfactory immunocytochemical reactions. For large or heavily mineralized cortical bone specimens, strong acid is used for decalcification but must be monitored by decalcification end point test.

B. *Weak Formic Acid*

Alternatively, formic acid, a weak acid, is used for decalcification. To counteract the injurious effects of acid, buffered formic acid is used and time is needed for complete decalcification. Below is the formula for buffered formic acid (Evans and Krajian, 1930)

20 % aqueous sodium citrate - 65 ml
90 % stock formic acid - 35 ml
The mixture should give a final pH of 2 to 3.

Small thin slabs of bone require two to four days in this decalcifying fluid. Small pieces of cancellous bone take 24 to 48 hours. Before processing, tissues are washed overnight in running water.

C. *EDTA*

Diaminotetracetic acid can also be used as a chelating agent for decalcification. It is an excellent bone decalcifier for immunohistochemical or enzyme staining and electron microscopy. It does not damage tissues or their stainability if buffered to neutral pH of 7 to 7.4. Decalcifying of dense cortical bone may take six to eight weeks. For small bone specimen, decalcification may be less than a week. After decalicification is completed, the bone tissue is transferred directly to 70% alcohol. The formula is given below:

EDTA, disodium salt - 250 g
Distilled water - 1750 ml
Neutralise to pH 7 with approximately 25 g of sodium hydroxide

Tissue Processing

Tissue requires dehydration through increasing strengths of alcohol before paraffin wax impregnation. Each stage of processing must be of sufficient length to ensure complete dehydration. Tissues must always be of a suitable size.

Tissue Tek cassette is commonly used to contain the sample. Do not squeeze large blocks of tissue to fit the cassette. This causes distortion and poor processing of tissue. Most laboratories use automatic tissue processing machine (Fig 4.2.1), which does not need any attention until the end of the sequence when the tissues are wax-impregnated prior to embedding.

After wax-impregnation the tissue is turned into a block with molten wax in an embedding center, which comprises of wax dispenser, cold plate and a heated storage for tissue moulding. Tissue block is then ready for sectioning after paraffin wax has solidified.

An overnight schedule is used to process soft tissue in our laboratory. Table 4.2.2 below provide time frame requirement for each processing step

Table 4.2.2: Time frame requirement for tissue processing

Container nos	Chemical Solution	Time (hours)
1	10% Formalin	0
2	70% Alcohol	0.5
3	95% Alcohol	0.5
4	95% Alcohol	0.5
5	100% Alcohol	1
6	100% Alcohol	1
7	100% Alcohol	1
8	100% Alcohol/Xylene (equal volume)	1
9	Xylene	1
10	Xylene	1
11	Wax	2

Fig 4.2.1: Tissue processor.

Microtomy

Sections of wax-embedded tissue are cut using a microtome (Fig 4.2.2) so that examination by microscopy can take place. Thickness of section ranges from 3 to 5 μm. The thickness of the section depends on the texture of tissue. Thin sections with crisp cytologic detail are essential for the diagnosis of haematopoietic and lymphoreticular neoplasms. Generally thicker sections are required of tissues such as normal lung and thinner sections of dense cellular tissues such as spleen. Hotplate is used for drying the cut sections. Drying takes five minutes. Do not overheat the sections as it will distort the tissues especially collagen. Overnight heating at 37 °C is recommended for immunohistochemistry staining.

Histological Staining

Various types of staining processes are employed to give good visualization, differentiation of cells and cellular details. Haematoxylin and eosin stain is widely used and serves as the main diagnostic technique in histology. Haematoxylin stains the cell nuclei blue/black and gives a good nuclear detail. Cytoplasm of cell will be stained by eosin while others such as connective tissue fibres will vary from intensities of pink, orange and red. Step-by-step method in haematoxylin and eosin staining technique is given below:

1. dewax sections, hydrate through graded alcohol to water
2. stain with Harris haematoxylin solution (Harris 1900) for 10 minutes
3. wash well in water
4. differentiate in 1 % acid alcohol
5. wash well in tap water

6. blue by dipping in ammonia water
7. wash well in running tap water for 5 minutes
8. stain in 1 % eosin Y for 30 second
9. dehydrate through alcohols, clear and mount.

Different colourations and shadings of the cellular materials are given in Table 4.2.3.

Table 4.2.3: A list of different colours observed for different cellular material

Cellular Material	Colouration/Shading
Nuclei, cytoplasmic RNA, some calcium salts	blue/black
Cytoplasm	varying shades of pink
Muscle fibres, keratin, coarse fibres , fibrin, fibrinoid	bright red
Red blood cells	orange/red
Collagen, reticulin nerve fibres, amyloid	pink

Selection of Special Stains

A. *Connective Tissue*

Masson Trichrome (Masson, 1929)

This stain is used for the differential demonstration of connective tissues. It gives a clear-cut differentiation of muscle from collagen, collagen fibres, fibrin and erythrocytes.

Connective tissue - *Blue oar green according to the counterstain use*
Collagen - *Blue/green*

Verhöeff Elastic Van Gieson (1908)

Elastic fibres vary in size and diameter. They are well seen in tissue sites in dermis of skin, lung, heart and blood vessel walls. It is a popular method and excellent result is seen when Van Gieson is used as counterstain.

Nuclei and elastic fibres - *Black*
Collagen - *Red*
Other structures - *Yellow*

Gordon & Sweets Reticulin (1936)

Reticulin fibres are difficult to see in Haematoxylin & Eosin (H & E) preparation. They are normally best seen in lymphoid tissue and liver. In myelosclerosis, there is an excess of reticulin fibres.

Reticular fibres - *Black*
Collagen - *Yellow/brown*
Background - *Clear if not counterstained; Red if counterstained*

B. Undecalcified Bone

Goldner's Stain (1937)

It is the most popular trichrome stain for undecalcified bone. The osteoblast and osteoclast activities are easily assessed. Osteoid appears red and mineralized bone green which gives a pleasant contrast. This stain gives a clear nuclear and cytoplasmic staining, a prerequisite for differentiation of bone and marrow cells. Other tissue components such as collagen, muscle and epithelia are also clearly defined.

Osteoid	*- Orange/red*
Mineralized bone	*- Green*
Nuclei	*- Blue/gray*
Cartilage	*- Purple*

Movat Pentachrome Stain (Olah *et al.*, 1977)

It is by far superior to Goldner's stain and designed for simultaneous demonstration of different fibres and ground substances in connective tissue. In plastic embedded undecalcified tissue, osteoids stain deep red in colour and mineralized bone brightly yellow.

Nuclei	*- Bluish grey/black*
Collagen fibres, cartilage, mineralized bone	*- Yellow*
Osteoid, muscle, cytoplasm, elastic fibres	*- Red*
Calcified cartilage	*- Sea green*
Ground substance mucin	*- Blue*

Von Kossa Reaction for Calcium Phosphate (Krutsay, 1963)

Von Kossa (Von Kossa, 1901) is the method used for detecting calcified matrices in bone. The calcified tissue components stain in various shades from brown to deep black. Modified method published by Krutsay improves the results in a homogeneous blackening of all calcium compartmets. When it is counterstained with haematoxylin (Ponceau acid-fuchsin), it yields a better result.

Calcium	*- Black*
Osteoid	*- Dense red*

Periodic Acid Schiff's PAS Combined with Alcian Blue (Hotchkiss, 1948)

Polysaccharides and proteoglycan are best demonstrated by periodic acid shiff reaction combined with alcian blue. This method is in fact an almost classical procedure for the study ossification as occurs in growth plates. The alcian blue is used in a slightly acid solution and binds preferentially to proteoglycans within the pericellular matrix and in calcified cartilage.

Nuclei, proteoglycan	*- Blue*
Cartilage, calcified bone	*- Pink*
Osteiod	*- Slight pink*
Glycogen	*- Dark red*

C. *Neuropathology Tissue*

Modified Bielschowsky's Silver Stain (Chan, unpublished)

This modified method stain for neurofibrils, dendrites and axons. It is particularly good in showing small groups of neurons bedded in deep nuclei of the central nervous system.

Neurofibrils, dendrites and axon	*- Black*

Plamgren's Stain (1948)

Different silver impregnation methods are used in staining the peripheral nervous system. This modified method is used to suppress staining of collagen and reticulin, which do not present a problem in methods destined for use in the central nervous system.

Nerve fibres	*- Brown or black*

Luxol Fast Blue (Klurer & Barrera, 1953)

Luxol fast blue is a copper phthalocyanine dye and accoding to Pearse (1955) has a strong affinity for phospholipids and strong choline bases. It is a popular method routinely used in fixed, paraffined processed sections for myelin. Counterstained nuclear fast red shows degenerate myelin in a contrasting colour.

Red blood cells, myelin	*- Blue to purple depending on the amount of counterstaining remaining*
Nuclei and Nissle substance background	*- Colourless*

Fig 4.2.2: Microtome.

CHAPTER 4.3

ANIMAL TISSUE PERFUSION AND PRESERVATION

Robert Ng

Liver

Perfusion of the liver is done during liver transplantation surgery and for harvesting of hepatocytes for culture. The former is a survival surgery involving cold ischaemic preservation of viable liver organ for allograft transplant while the latter is a nonsurvival procedure. Perfusion is based on the portal vascular system. This chapter will focus on harvesting of hepatocytes from liver organ for research use.

The anaesthetized animal is first given heparin (150 IU/kg) prior to perfusion. The portal vein is mobilized and a Bardec catheter is used for cannulation, paying special attention to ensure that air is not introduced into the flow system. This can be achieved by filling the catheter with cold Hartman's solution and clamping both ends with tubing clamps. The superior vena cava and hepatic artery are mobilized, ligated and excised. The liver and the attached vessels are lifted off its bed after cauterizing all leaked vascular points and then transfer to a perfusion tray. The portal Bardec catheter is connected to the occlusion pump for pulsating flow at 400 ml/minute for pig of size 15 to 20 kg. For rat liver, perfusion can be done *in situ* using a 50 ml syringe pump or by gravity flow from a height of six feet above the perfused liver using a flow rate of 10 to 15 ml/minute. A slit is made on the inferior vena cava for venous retrograde outflow of the perfusate.

The first perfusate used is cold Hartman's solution to preserve the organ and remove blood until it is completely blanched. This is then followed up with 0.1 % collagenase D in phosphate-buffered saline (PBS) and liver is perfused for 10 to 20 minutes until the liver tissue appears soft, fragmented and spongy as indication of sufficient digestion and cellular detachment. The tissue is gently mashed on a 100 μm sieve tray together with jet spraying of William E culture medium with 10 % foetal calf serum to deactivate the activity of collagenese, which would otherwise cause cellular death on prolonged exposure. The cell suspension is collected and centrifuged at 900 rpm for 5 minutes to pellet the cell, which is tested for viability using trypan blue dye occlusion test before being cultured.

Brain and Eye

Vascular perfusion of the brain and eye with formalin and glutaraldehyde allow for quick fixation of tissue before autolysis sets in. This process involves a two-stage approach. The deeply anaesthetized pig (30 kg weight) is first injected with heparin (150 IU/kg) intravenously. The chest cavity is opened. The left ventricle is cannulated with a trocar needle (3 mm diameter) attached to a 3 mm diameter tubing that is connected to an occlusion pump for pulsating flow rate of 300 ml/minute and ensuring no air is inducted during the perfusion process. Two and a half litres of cold Hartman's solution are required to preserve the tissue and drain off the blood. The right atrium is nicked for perfusate outflow discharge and the descending aorta is clamped to confine perfusion mainly to the head and neck region. This is followed with perfusion of either 10 % formalin or 2.5 % glutaraldehyde to fix the tissue efficiently. Circumferential craniotomy is performed to harvest the whole brain organ. Similarly the eye can be harvested using a scalpel blade and scissors to navigate and cut around the socket to facilitate detachment of the eye with its optic nerve. Better result can be achieved if the vitreous is replaced with fixative to retain the eyeball architecture.

Abdominal Organs

For generalized body perfusion of the abdominal organs, similar perfusion steps for brain and eye can be used except that the aorta is not clamped. Alternatively, the femoral or carotid artery can be cannulated for inflow of perfusate and the contralateral vein is nicked for venous discharge. 10 % formalin is normally used as perfusate. However, for better tissue preservation 2.5 % glutaraldehyde or 3.0 % paraformaldehyde in phosphate-buffered saline (PBS) is preferred but the choice is based on study requirement. The target organs are harvested, slice if required and stored in 20 volume of 10 % formalin to fix tissue overnight in the refrigerator before being processed for histology.

Cold Preservation

For RNA work or toxicogenomic microarray studies, it is recommended that rodents be humanely euthanized with decapitation as the method is likely to induce less RNA changes. Tissue harvesting should preferably be within two minutes from time of death with different personnel tasked for separate harvesting of tissue from the head and the abdomen. The organs/tissues collected will include frontal cortex, cerebellum, brain stem, hippocampus, right/left retina, right/left testis, skeletal muscle (thigh), pancreas, brown fat (around kidney), white fat (mesenteric fat), right/left kidney, liver (median lobe), small intestine (duodenum, jejunum and ileum), left lung (anterior portion), right lung (apical lobe) and heart (left ventricle). Tissue slices are collected with aluminium foil boat, which is pre-cold in liquid nitrogen. The aluminium foil tissue is then snap frozen and deposited in cryogenic screw cap tube or 50 ml Falcon tube and immediately transferred to –80 °C freezer.

To preserve cells, they are suspended in 1.0 ml of freezing medium (10 % dimethylsulfoxide, DMSO in foetal bovine serum, FBS) in a cryogenic vial. The vial cap must be tightened securely without dislodging the rubber gasket for leak proof assurance. The vial is first transferred to a –80 °C freezer for 6 hours or overnight before finally depositing the vial into a cryobox for deep freezing in a liquid nitrogen cryogenic chamber.

CHAPTER **4.4**

ANIMAL CELL CULTURE

Peggy Yong and Kai Zhang

Cell culture is an important tool in biomedical research. It was first successfully performed in the early 1900s. From late 1940s onwards, three crucial developments, as listed below, had made cell culture a more viable tool for scientists:

- Development of antibiotics that are subsequently used as medium requirement on most cell lines, which avoided contamination often encountered in earlier attempts in cell culture.
- Development of other techniques, such as the use of Trypsin-EDTA, which is used as a cell detachment buffer. This step is crucial, allowing for cell harvesting to propagate the cell lines.
- Development of standardized, defined culture media that allows for continuous growth of cell.

Generally, normal tissues will give rise to a finite cell line that is the cells will die after several subcultures. However, a finite cell line may be capable of transforming into a continuous cell line. On the other hand, cells from malignant tumours may transformed to give rise to a continuous cell line and are easier to grown albeit the disadvantages of the loss of donor phenotype, chromosomal stability and tissue specific markers.

Nowadays, most cell lines are propagated as monolayer cultures, for example HepG2 cells (Fig 4.4.1), adhered onto a plastic or a glass surface. Some cultures, such as ascites tumours can be propagated in suspension. These cultures are grown in controlled environment, both within the flask and the incubators used.

The general principles underlying cell culture techniques and precautions are detailed below.

General Principles of Cell Culture

The main purpose of understanding these principles is to allow researchers an insight into cell culture techniques, though this is not necessarily applicable to all types of cell lines.

1. *Cell Adhesion to Culture Flask Surface*
   The propensity of cultured cells to adhere to culture surfaces is fundamental to monolayer cell culture technique. Therefore, it is important to use techniques that will enhance the adhesion process. The culture surface must be hydrophilic and correctly charged before adhesion of cells can occur. Cells possess unevenly distributed negative surface charges and can be cultured on either negatively-charged surfaces such as glass and plastic or positively-charged polylysine-coated surfaces. Since cells can adhere and grow on all these surfaces, the main factor governing adhesion is the density of the charges rather than the polarity of the charges.

   The media used in culturing the cells also play important role in cell adhesion to the culture flask surface. The media used contain divalent cations and proteins that allow for specific adhesion to the surface. Either cold insoluble globulin or fibronectin must be absorbed on the culture surface before it can promote cell attachment and spreading. Culture medium supplemented with 10 % (v/v) foetal bovine serum contains about 2 to 3 ug/ml of fibronectin, which absorbs onto the culture surfaces within a few minutes, will help to promote cell attachment and spreading.

2. *Choice of Culture Media*
   It is important that culture media have a large reserve of essential nutrients to support the growth of cells. Each of the various components in the culture medium has a contributory part to play in sustaining and promoting cell growth.

   - **Amino acids**
     Amino acids such as cysteine, glutamine, isoleucine and serine are the amino acids utilised most rapidly in cell cultures. Deficiencies in supply of any one of the essential amino acids stress culture cells and retard cell division, causing chromosomal damage and increase lysosomal activity.

   - **Nucleic acid precursors**
     A supply of components such as adenosine, guanosine, cytidine, uridine and thymidine is often beneficial particularly in cases where folic acid is in short supply and when cells are cultured in low densities.

   - **Carbon sources**
     Growth of cells in culture depends on carbon sources. In most of the commonly used culture media, this source is provided by glucose (5 to 20 mM) and glutamine (1 to 5 mM). Media containing glucose should be supplemented with pyruvate (1 mM), for the growth of cells under conditions of low density. The type of carbon source in culture media influences the formation of lactate.

- **Vitamins and choline**
  Additions of retinoids can promote adhesion of cells, by influencing the synthesis of specific glycoproteins on the cell surfaces. Addition of choline is important when using low concentrations of serum (less than 5 %).

- **Polymers**
  A high molecular weight component may be necessary for culture of some types of cells and at low cell densities. When working with cells possessing very low plating densities (< 1 %) and at low inoculation densities, it is beneficial, during the early stages after inoculation of the culture, to include a polymer in the media such as Dextran T-D, Dextran T-500 or methylcellulose. If urea cycle compounds and other products of metabolism accumulate to toxic levels at later stages of the culture cycle and replacement of medium is not possible, carboxymethylcellulose can be added to reduce the effects of these toxic components.

3. *Serum Supplement*
In the absence of serum, most cells fail to proliferate. Hence, serum supplement is an essential component of culture media. Sera used to supplement culture media come from a variety of sources and are used at concentrations ranging from 0.5 % to 30 % (v/v).

The serum serves two vital functions. Firstly, it assists attachment of cells to the culture surface most likely by supplying exogenous glycoproteins involved in the attachment process. Presence of growth factors and hormones in serum also promotes proliferation of the cells.

Besides these two vital functions, serum has a protective effect on cells in culture and hence enhances viability. During routine subculture procedures, serum provides protease inhibitors, which inactivate trypsin used to detach the cells from culture flasks. Cells left in trypsin for a long period may affect the viability of the cells. Therefore, addition of serum-supplemented medium will prevent the cells from dying.

4. *Gas Supply*
The supply of the correct amount of O_2 and CO_2 is important in achieving high yields. Both these gases have important metabolic functions and CO_2 is also involved in the control of culture pH.

Monolayer cultures are relatively anaerobic and many established cell lines commonly used are adapted to such conditions. The partial pressure of O_2 in body fluids is less than that of air and although more culture media have been developed for use in about 20 % CO_2, the tension optimal for cell growth is often substantially lower. In the laboratory, it is usually sufficient to use 95 % air as source of O_2, with 5 % CO_2, throughout the culture cycle.

Solution of CO_2 in media results in the formation of HCO_3^-, which is intimately related to pH. It has been difficult to define the tension of CO_2 optimal for cell growth. 5 % CO_2 concentration was selected originally on the basis of its being the concentration in the alveolar spaces of the lungs. This concentration was intended for general cell culture. Certain cell lines may not need CO_2, for example MDA-MD-231 cells (Fig 4.4.2). This cell line grows in Leibovitz L-15 medium, which does not rely on CO_2 for buffering and control of pH. This medium is also used when low tension of CO_2 is required.

5. *Culture pH Maintenance*

Since pH influences cell survival, attachment, growth and function, maintaining the correct pH is crucial to obtaining optimal cell growth and high yields. Methods for controlling pH include buffering to minimize the effects of lactate on culture pH or altering culture conditions such that the cells produce less lactate. The pH of cell culture medium is monitored with phenol red, an indicator for change of pH in the culture medium.

Most media utilize a CO_2/HCO_3^- buffer system, which is not sufficient to prevent a decrease in pH towards the end of the culture cycle. If a cell type tends to produce large quantities of lactate, then a formulation with a higher concentration of HCO_3^- should be used.

Addition of HEPES delays the onset of pH drift and usually increases cell yield. HEPES assists in maintaining pH during the attachment period of culture and plating efficiency is enhanced. Whenever the tension of CO_2 is low, there is a lower stability of the HCO_2^- system and HEPES should be used. The exact amount of HEPES used should not be more than what is required to maintain the pH and it is advisable to start with 10 mM.

Some established cell lines produce large quantities of lactate during cell metabolism especially at the tail-end of the culture cycle causing a rapid decrease in pH. Since culture cells degrade glucose to either CO_2 or lactic acid, high concentration of glucose results in the formation of high levels of lactate. To reduce the production of lactate, glutamate should b used as an alternative source of energy because its degradation to lactate is less than in the case of glucose. The amount of lactate secreted by transformed cells can also be reduced by Biotin and replenishment media can be supplemented with this component. Another method of reducing lactate formation is to use galactose (2 to 10 mM) as a carbohydrate substitute for glucose as in Leibovitz L-15 medium.

6. *Cell Harvest*

When the cell growth on the culture flask has reached 80 to 90 % confluency, the cells will be harvested and subcultured. A detachment buffer, one of which that is often used is Trypsin-EDTA, is needed for this process.

The success of harvesting with trypsin is dependent on ensuring that harvesting is done with the medium pH between 7.4 and 8.0 since cells are less strongly attached at alkaline pH. The combination of chelating agent (EDTA) and a proteolytic enzyme (trypsin) is especially useful for cells where attachment is very strong. Detachment of cells from culture surfaces is monitored and once completed, trypsin has to be deactivated by adding culture medium containing 10 % serum (v/v).

7. *Cell Storage*
   Harvested cells are resuspended in a specially concocted medium to a density of 1×10^6 cells/ml. The most popular concoction is 10 % dimethylsulfoxide (DMSO) in foetal bovine serum.

   The resuspended cells are stored in special cryotubes. It is very important to gradually freeze the cells to maintain viability of the cells. It is recommended in lowering the temperature at a rate of 1 °C/minute until it reaches –80 °C. After which the cells will be kept for at least eight hours in –80 °C freezer before being stored in the liquid nitrogen tank (–150 °C).

Cell Culture Sterile Techniques

The cell culture technique described below is for serum supplemented cell culture in flask. Sterile techniques have to be maintained to avoid contamination of cultures. Some general sterile conditions are listed below:

1. All liquid used such as phosphate buffered saline and trypsin, have to be sterile. If the solutions are prepared in-house such as the case for phosphate buffered saline, the solution has to be autoclaved and deionised water to be used.
2. Switch on the ultraviolet lamp of the laminar chamber for at least half an hour before proceeding with any culture.
3. All necessary solutions and media are warmed to 37 °C.
4. Turn off the ultraviolet lamp and switch on the light and the airflow. Clean the hood with 70 % alcohol and bring all prewarmed items into the hood. The following reagents are frequently needed in cell culture.

   - Culture medium that is suitable for the cell line. The culture medium is supplemented with 10 % foetal bovine serum and 1 % antibiotics. In some cases, non-essential amino acids (NEAA) and L-glutamine are also added.
   - Buffer solution – most commonly used buffer solutions are Phosphate Buffered Saline (PBS) and Hanks' Balanced Salt Solution (HBSS).
   - Cell detachment buffer – most commonly used is 1 × Trypsin-EDTA (0.25 % trypsin).

Below are general descriptions of cell culture techniques such as culturing cells from frozen stock, subculturing of adherent cells, cell viability count and thawing cells from frozen stocks.

1. *Cell Thawing from Frozen Stock*

   - A vial of cells is taken from the liquid nitrogen tank.
   - The cells are thawed in 37 °C waterbath for one minute or until the cells have completely thawed.
   - Clean the surface of the vial with alcohol and put in into the culture hood.
   - Place 19 ml of cell culture medium into a 75 cm^3 flask.

- Seed all cells into the flask and properly label the flask. Labeling of the flask will include information such as the date of thawing and type of cells.
- Place the flask into a suitable 37 °C CO_2 incubator. Some cell lines require non-CO_2 incubator.
- The following day, observe the cells under the microscope. If most cells have adhered to the flask, the medium is changed. This step is important because the presence of DMSO in the frozen stock can be detrimental to the cell viability.

2. *Subculture of Adherent Cells*

- Prepare a waste bottle, which contains about 10 to 20 ml of concentrated bleach.
- Discard the used culture medium from the culture flask into the waste bottle.
- Wash the cells with 10 ml of PBS. This step is repeated for a second time.
- Add 2 ml of trypsin-EDTA to the flask and incubate or until the cells have properly detached.
- Gently rock the flask and observe the cells under the microscope. Once the cells have rounded up and floating, the detachment process is considered complete.
- Add 8 ml of serum-supplemented medium to the flask.
- Draw the suspended cells into a 15 tube.
- Depending on preference and speed of cell growth, a suitable subculture ratio is used. As a general guidance, seed 1×10^6 cells into a 75 cm^3 culture flask containing 19 ml of culture medium.
- Properly label the culture flask with the cell type, culture ratio, date of subculture and passage number.

3. *Cell Viability Count*

Very often, trypan blue is used to stain the cells. Viable cells will exclude the dye but dead cells will take up the dye. However, if exposed to the stain for extended period, both viable and dead cells may take up the dye.

- Follow the same methods as for subculturing cells.
- After drawing the cells into a 15 ml tube, centrifuge the cells at 1500 rpm for 5 minutes.
- Carefully decant the supernatant and resuspend the cells with 5 ml of buffer solution.
- Centrifuge the cells again at 1500 rpm for 5 minutes.
- Decant the supernatant and resuspend the cells in 5 ml of medium of choice.
- Take 10 ul of the resuspended cells and add to 10 ul of trypan blue stain. This will give rise to dilution factor of two. Pipette up and down to allow for thorough mixture. Leave it to stand for 5 to 10 minutes.
- Add 10 ul of cells to the haematocytometer and count viable cells under the microscope.
- Cells per square should be about 30 to 50 cells. If the cell count is too high, redilute to an appropriate dilution factor.

4. *Cell Freezing*

- The freezing solution is prepared. Usually 10 % DMSO in foetal bovine serum is used.
- The cells are harvested using trypsin as in subculturing.
- Resuspend the cells in the chosen freezing solution to a concentration of 1×10^6 cells per ml. Distribute 1 ml into each labeled cryotubes. Transfer vial to –20°C for 1 hour, –80 °C for 8 hours and finally in liquid nitrogen freezer (–150 °C).

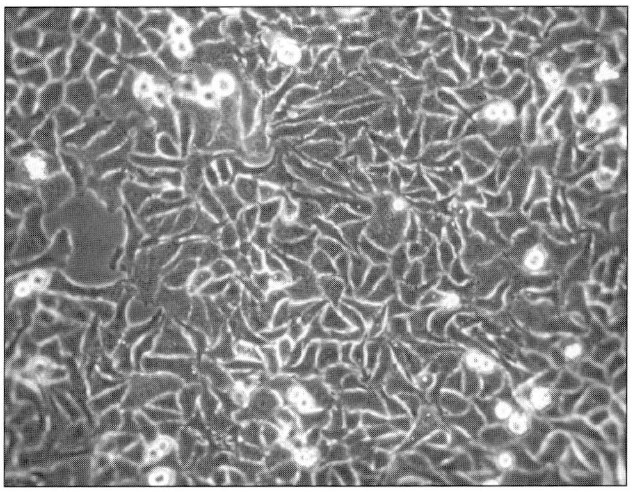

Fig 4.4.1: HepG2 cells observed under the microscope.

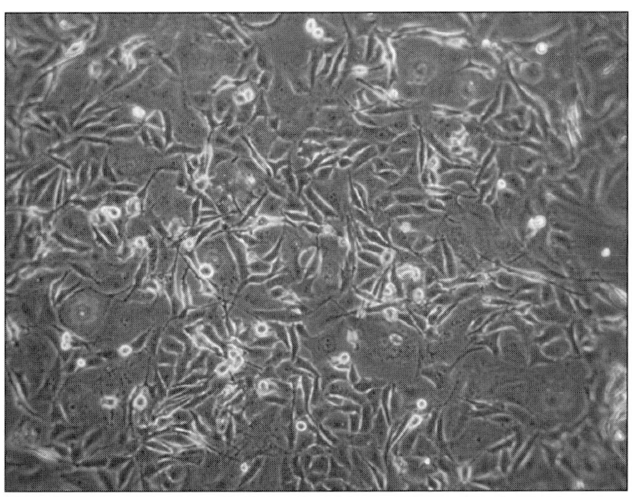

Fig 4.4.2: MDA-MB-231 cells observed under the microscope.

CHAPTER **4.5**

APPLICATION OF MICROSURGICAL TECHNIQUES IN ANIMAL RESEARCH

Bien Keem Tan, Evan Woo, Colin Tham, In Chin Song, Angela Goh and Bien Soo Tan

Microsurgery means surgery with the aid of a microscope, and in the field of plastic surgery, it encompasses microvascular surgery, microneural surgery, microlymphatic surgery, and microtubular surgery. Microvascular surgery refers to coaptation of small vessels performed under illumination and magnification. Animal research that involves microsurgical techniques include microcirculation studies, flap physiology studies, flap prefabrication and tissue transplantation, to name but a few. The purpose of this presentation is to share our experiences of applying microsurgery in animal research.

Microcirculation Studies

These may be classified into static and dynamic studies. Static studies of the vascular pattern of tissues or organs may be obtained by dye injection studies, corrosion casting and radiography (Figs 4.5.1 to 4.5.7). Dynamic studies include microvascular video imaging that allow image enhancement, documentation of blood flow and quantification of capillary fluid exchange.

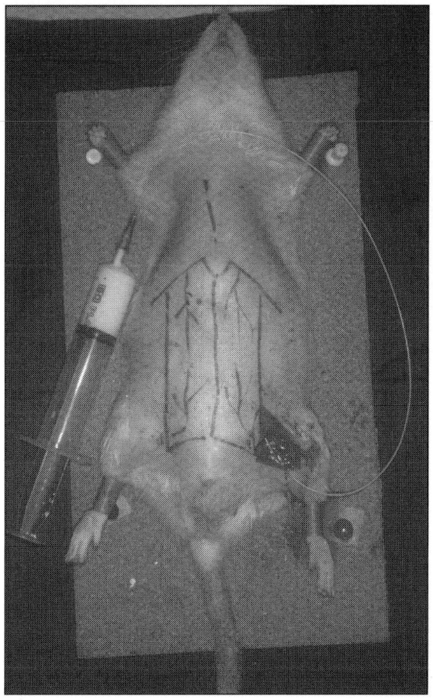

Fig 4.5.1: Injection of barium sulphate-gelatin suspension to study vascular anatomy in the rat.

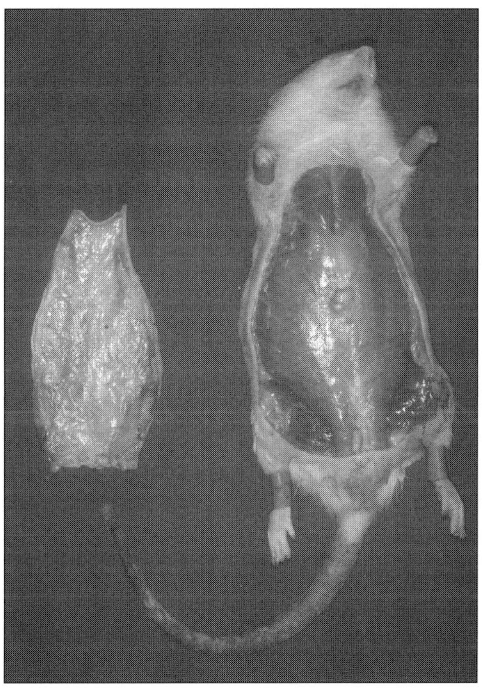

Fig 4.5.2: Rat ventral skin harvested after injection.

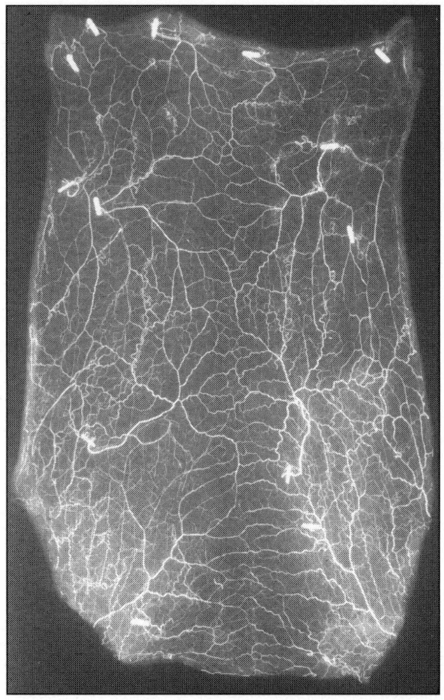

Fig 4.5.3: Vascular pattern of the ventral skin demonstrated by soft X-rays.

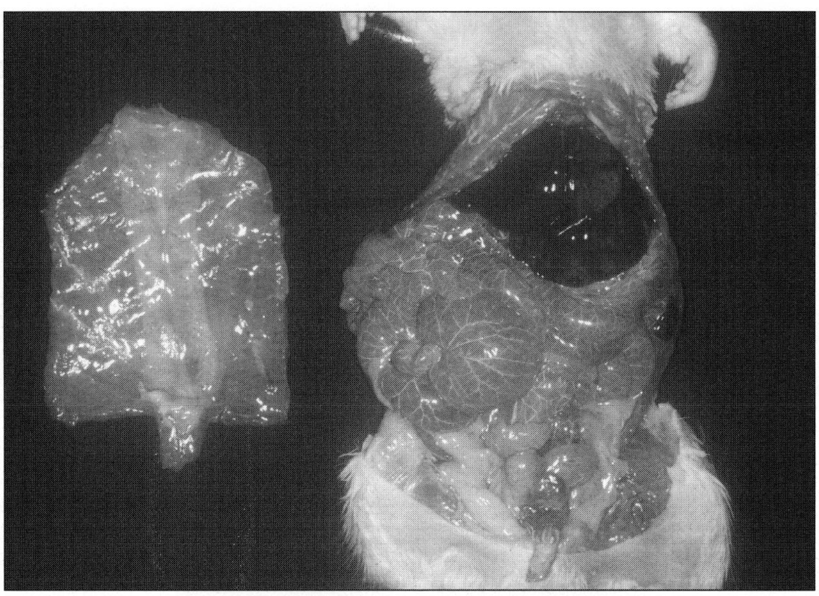

Fig 4.5.4: Rat ventral abdominal wall.

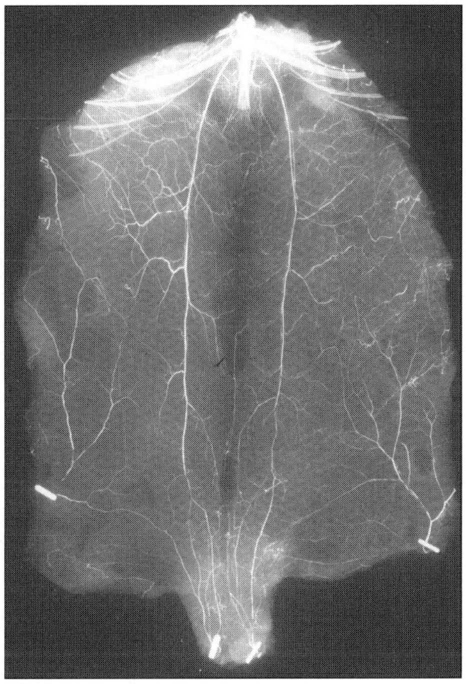

Fig 4.5.5: Microangiograph demonstrating the paired recti muscles supplied by the deep inferior and superior epigastric arteries.

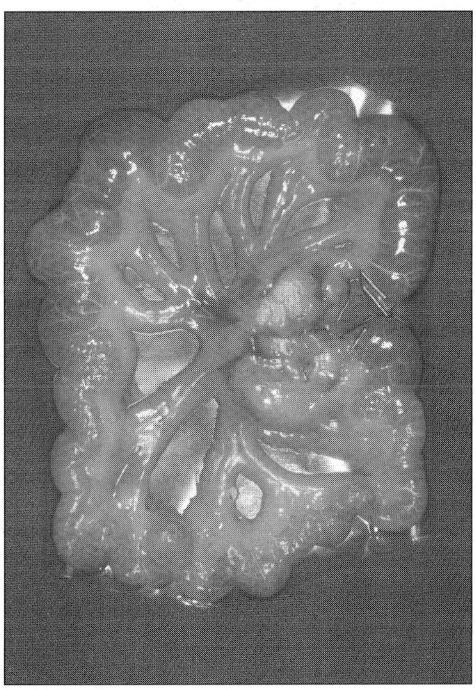

Fig 4.5.6: Rat intestine.

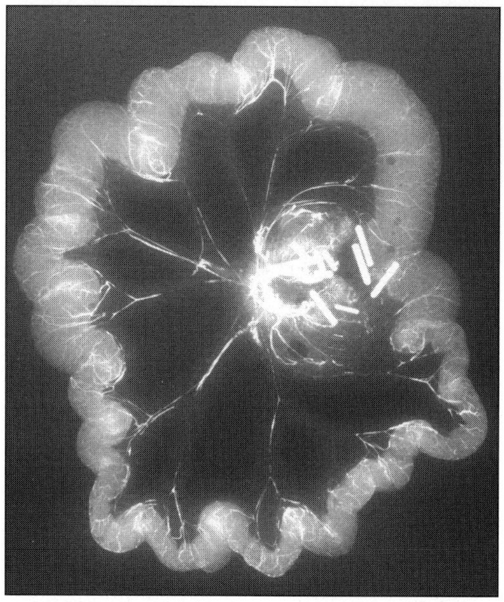

Fig 4.5.7: Microangiograph showing intestinal blood supply.

Flap Prefabrication

Flap prefabrication is one of the most exciting areas in plastic surgery because of its bridging role between conventional reconstructive surgery and tissue engineering. Using this technique, tissues such as bone, cartilage, skin and muscle can be pre-assembled to form precise composites that will fit any defect (Khouri R. K., Upton J., Shaw W. W., 1991; Shintomi Y. and Ohura T., 1982; Erol O. O. and Spira M., 1980; Erol O. O., 1976; Morrison W. A., Dvir E., Doi K. *et al.*, 1990). In its simplest form, prelamination, for example, a ear may be created by burying cartilage underneath the forearm skin and later harvested as a skin-cartilage composite free flap to replace the missing part.

Another method, vascular induction, is one in which new blood supply is introduced to create new transplantable tissue. For example, bone chips wrapped in a vascular carrier such as muscle become vascularized grafts (Fisher J. and Wood M., 1987). The example below describes jejunal prefabrication in a rat model using the same technique (Tan B. K., Chen H. C., Wei F. C. *et al.*, 2002). Intestinal segments wrapped in muscle flaps become independent of their mesenteric blood supply by "parasitizing" on the muscle's blood supply.

This idea arose from our initial observations that intestinal segments transferred to the neck to reconstruct the oesophagus could survive accidental disruption of the pedicle if sufficient time had elapsed (Chen H. C., Tan B. K., Cheng M. H. *et al.*, 2002). Clearly, the bowel had picked up new blood supply from its bed. With this observation, could we innovate a technique to deal with difficult cases of oesophageal reconstruction, in which recipient vessels for free jejunal transfer are often unavailable because of previous surgery or irradiation? In such instances, jejunal transfer employing a muscle flap as a "vascular carrier" may provide a solution to the problem. Could jejunum be induced to acquire a new blood supply from a muscle flap wrapped around it? If so, the bowel could be carried by the pedicled muscle flap up to the neck for oesophageal reconstruction.

1. *Intestinal Prefabrication*

The present study was designed to evaluate whether small bowel could be vascularized by a muscle flap for survival without mesenteric blood supply (Tan B. K., Chen H. C., Wei F. C. *et al.*, 2002). Muscle was selected as the vascular carrier because it is commonly used in head and neck reconstruction. The time required for successful revascularization was evaluated by mesenteric ligation at predetermined time intervals. The mechanism and quality of revascularization was assessed by histology and microangiography.

Twenty-four mature (500 to 700 g) rats were divided into six experimental groups of four animals each. In each animal, a 1.5 cm segment of proximal jejunum was isolated on two jejunal arteries and wrapped around with a superior pedicled rectus abdominis muscle flap (Fig 4.5.8). To determine the time of neovascular takeover, the mesenteric pedicles were ligated on postoperative Day 2 (Group I), Day 3 (Group II), Day 4 (Group III), Day 5 (Group IV), Day 6 (Group V) and Day 7 (Group VI). At the time of pedicle ligation, the composite flap was transposed to a new subcutaneous position. Viability of bowel was assessed according to gross appearance and histology 48 hours after transfer.

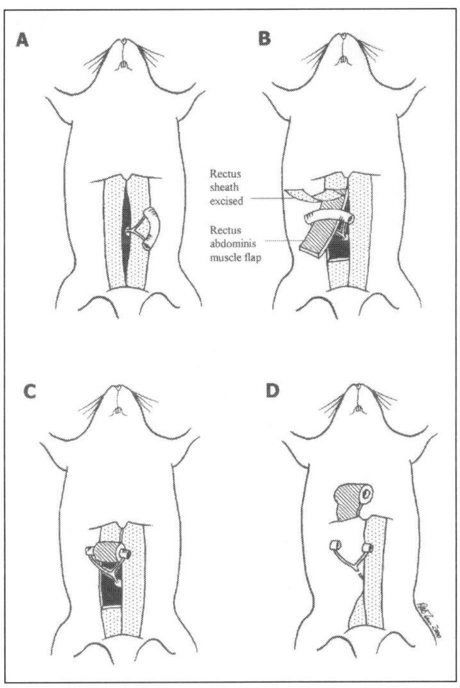

Fig 4.5.8: Schematic representation of jejunal prefabrication using the rectus abdominis muscle flab *(reproduced with permission from Academy of Medicine, Singapore)*.
(A) A 1.5 cm proximal jejunal segment based on 2 jejunal pedicles was isolated. (B) The anterior rectus sheath over the right rectus abdominis muscle was excised. (C) A superior-pedicled rectus abdominis muscle flap was elevated and folded around the bowel segment. (D) At the second stage, the mesenteric pedicle was ligated and divided. The bowel ends were cut flush with the muscle flap and the composite flap transposed to a new subcutaneous location.

Complete survival of revascularized jejunum in 11 of 12 animals was obtained after pedicle ligation on postoperative Day 5 and beyond ($p < 0.0001$, Fisher's Exact Test). These bowel segments demonstrated luminal patency, intact pink mucosa, mucous production and visible peristalsis (Fig 4.5.9). Histology showed healthy intestinal epithelium and tissue integration along the serosa-muscle interphase (Fig 4.5.10). Goblet cells, which secrete an acid mucous, were located along the crypts and villi. These stained pink with mucicarmine and had a normal distribution pattern. Alkaline phosphatase activity was demonstrated by red deposits along the brush borders.

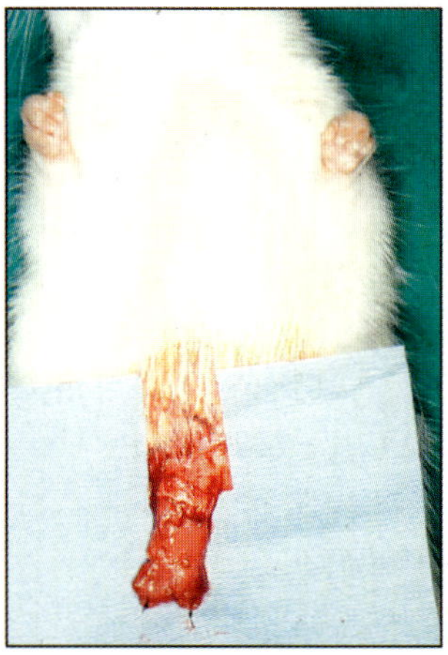

Fig 4.5.9: Prefabricated jejunal segment opened longitudinally to expose the mucosal surface *(reproduced with permission from Academy of Medicine, Singapore).*

The jejunal segment demonstrated patency of lumen and mucous production. It was lined with pink velvety mucosa coated with clear/cream-coloured viscous fluid characteristic of mucous. There was peristaltic activity, which could be elicited by stroking the mucosal surface with a cotton tip. The bowel was tightly adherent to muscle and did not separate easily with manipulation.

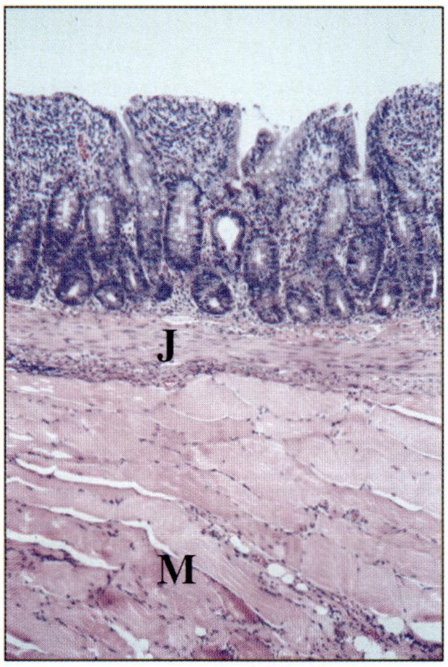

Fig 4.5.10: Histologic section of Group V prefabricated jejunum (mesenteric ligation on postoperative Day 6) *(reproduced with permission from Academy of Medicine, Singapore).*

Notice the intact jejunal mucosa characterized by tall villi and deep intestinal glands (crypts of Lieberkühn), festures typical of normal jejunum. These were lined by intact columnar epithelium interspersed with numerous plump goblet cells. At the muscle-bowel interphase, union of the skeletal muscle fibres with serosal was observed. Acellular gaps between the two surfaces were few. Note muscle (M) – jejunum (J) integration. Villous height and glandular distribution are compatible with normal jejunum; original magnification X 25.

In contrast, pedicle ligation on Day 4 and earlier resulted in varying degrees of bowel necrosis characterized by flattening or ulceration of mucosa (Day 4), mucosal sloughing and necrosis of mural musculature (Day 3), and complete loss of bowel architecture with lumen obliteration (Day 2).

This study demonstrated the feasibility of vascularising small bowel with muscle flaps to create bowel segments that are independent of their native mesenteric blood supply. Complete survival and normal function were evidenced by structural integrity, mucous production, enzyme secretion and peristalsis. The observations in this study indicate that the process of jejunal revascularization is time-dependent. In the present rat model, a minimal period of five days was required. This was statistically significant ($p < 0.0001$) as no viable bowel segment was found in animals that underwent mesenteric ligation four days and earlier. To explain the all or none survival of bowel, we postulate that Day 4 represented a critically ischaemic phase beyond which survival was assured if neovascularization remained uninterrupted. Short of this, all bowel segments underwent progressive autolysis resulting in necrosis.

Interestingly, it was observed that the quality of prefabricated jejunum improved with time. Microscopically, Day 7 and Day 6 specimens had taller villi and more numerous goblet cells than Day 5 specimens, although no discernible difference was seen on gross examination. One could postulate that with time, better neovascularization had promoted epithelial regeneration and growth.

Contact between muscle and bowel appeared an important factor in jejunal prefabrication. Microscopically, areas with good contact showed direct muscle to serosa healing, whereas areas with poor contact showed granulation tissue composed of fibroblasts, new blood vessels, loose extracellular matrix, and a variable collection of inflammatory infiltrate. Although the presence of granulations in the muscle-bowel interphase did not affect survival, we feel that these might predispose to scarring and consequently strictures in the long term. Hence, efforts should be made to maximize contact by using stitches or tissue adhesives such as fibrin glue.

2. *Clinical Application*

From a clinical standpoint, jejunal prefabrication may offer some problem-solving alternatives in difficult cases of oesophageal reconstruction. In instances where neck vessels are absent, jejunum transfer could be accomplished using a muscle carrier with an adequate reach. The potential for such an application was realized recently in a patient in whom neck vessels were unavailable because of irradiation (Chang S. Y., Chen H. C., Tan B. K. *et al.*, 1999). An exteriorized segment of jejunum was wrapped with a latissimus dorsi muscle flap for eight weeks before transfer. At maturity, the mesenteric pedicle was divided and the jejunum successfully transposed to the neck based on the muscle flap (Fig 4.5.11). With further experience, we should be able to speed up the process, using indicators such as goblet cell density and villous height as determinants of vascular sufficiency.

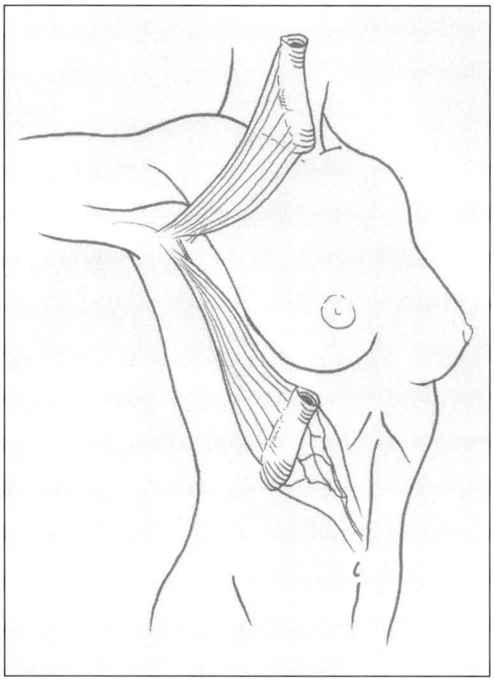

Fig 4.5.11: Oesophageal reconstruction in a patient by jejunal prefabrication using the latissimus dorsi muscle as a vascular carrier *(reproduced with permission from Academy of Medicine, Singapore).*
Reconstruction was achieved in two stages – eight weeks apart. First, to vascularize the intestinal segment with muscle and second, to transfer the composite flap to the neck to bypass the strictured oesophagus.

Second, in patients with diffuse intraperitoneal adhesions precluding dissection of the jejunal vessels, microsurgical jejunum transfer could still be attempted employing the muscle flap as a vascular carrier and pedicle.

Third, in patients undergoing long segment oesophageal reconstruction, flap prefabrication could be employed for lengthening the jejunal flap. Since bowel curvature is determined by mesenteric attachment, it is conceivable that bowel can be straightened by substituting the native curved mesentery with a straight muscle flap neomesentery.

The uses of muscle flap vascularization need not be confined to the gastrointestinal tract. Other organs, aside from bowel, may be revascularized in like manner. Recently, we salvaged a thrombosed live fibula graft used for tibia reconstruction by stripping away the periosteum and completely enveloping it with the tibialis posterior and flexor hallucis longus muscles (Figs 4.5.12 and 4.5.13). One year later, bone scans showed tracer activity (Fig 4.5.14) and the patient was able to walk with a brace (Fig 4.5.15).

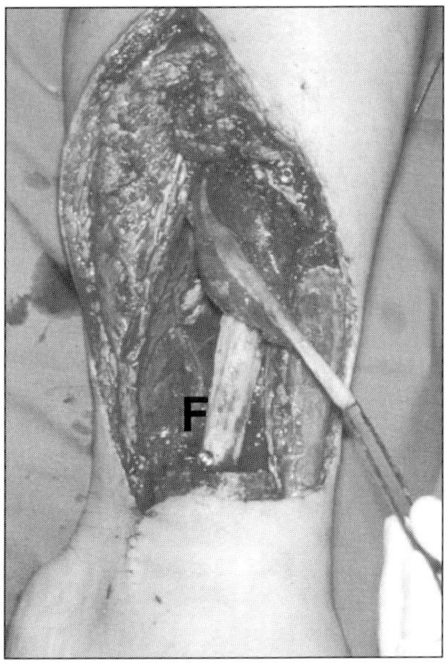

Fig 4.5.12: A nonvascularized fibula (F) was used to replace the tibia in a patient after tumour resection *(reproduced with permission from Academy of Medicine, Singapore).*

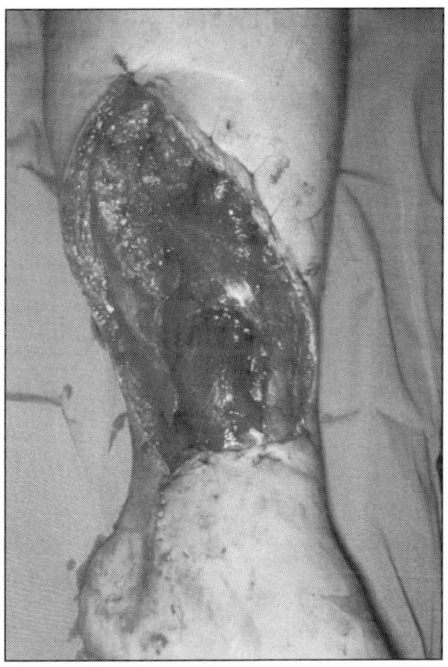

Fig 4.5.13: To ensure vascularity, the bone was completely wrapped with the tibialis posterior and flexor hallucis longus muscles *(reproduced with permission from Academy of Medicine, Singapore).*

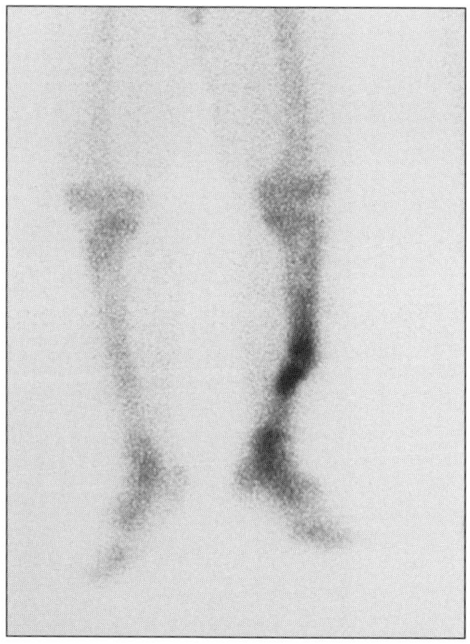

Fig 4.5.14: Bone scan of the same patient, showing tracer activity in the reconstructed segment ***(reproduced with permission from Academy of Medicine, Singapore).***

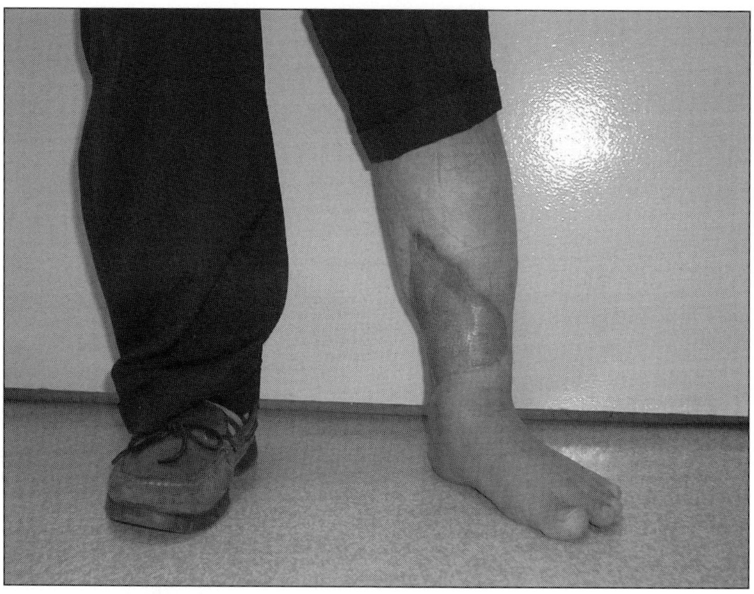

Fig 4.5.15: One-year follow-up assessment demonstrated the patient's ability to weight-bear and walk with a brace ***(reproduced with permission from Academy of Medicine, Singapore).***

3. *Pancreatic Prefabrication*
   Another application could be in the area of pancreatic transplantation, since present-day techniques have yet to overcome problems such as insufficient vascularity and unpredictable transplant survival. To transplant a portion of the pancreas, say its tail, the pancreas is first wrapped with a muscle flap, and then transplanted at a later stage as a prefabricated pancreatic flap. This would obviate the need for whole organ transplant, reducing donor sacrifice and morbidity. In this study, we investigated the feasibility of creating an independently functioning neo-pancreas from a portion of

pancreas from a rat. Our long-term aim was to develop this as a new technique to pancreatic transplantation.

Using a rat model, a neo-pancreas was prefabricated using a muscle flap to vascularize a segment of pancreas (Fig 4.5.16). The experiment was carried out in 3 phases. In Phase 1, a rectus flap was raised and juxtaposed to the tail of the pancreas to induce growth of blood vessels from muscle to pancreas. In the next phase, once the pancreatic segment was visualized by the muscle flap, it was separated from the rest of the gland. The blood supply of this segment of pancreas is now derived completely from the muscle flap and hence, may function as an independent neo-organ.

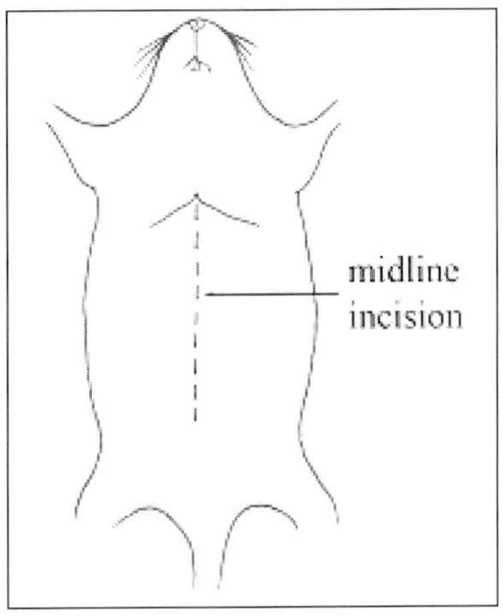

(A)

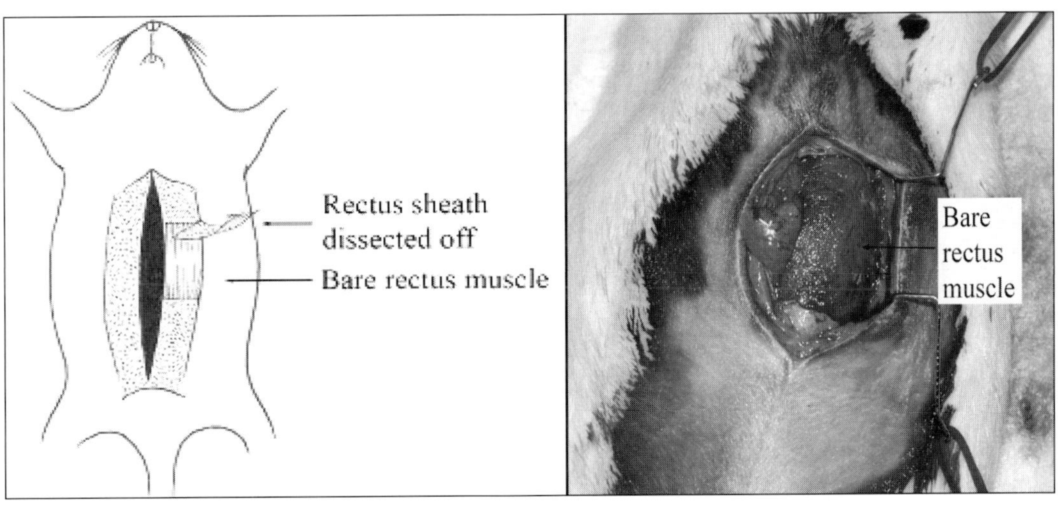

(B)

Fig 4.5.16: Illustration of surgical technique on rat.

(A) Dotted line showing midline incision. (B) Portion of rectus sheath is dissected off the rectus abdominis so that pancreas lies on bare muscle. (C) Close up of exploratory laparotomy showing pancreas and structures related to it. (D) The tail of the pancreas is mobilized and delivered extra-peritoneal to lie on bare muscle. The pancreas is sutured down to ensure that good contact with the muscle is maintained. (E) Abdominal closure is performed in layers from muscle to skin. Vascularization of the pancreas by the rectus muscle is expected after five days. (F) The extra-peritoneal pancreas is divided from the rest of the pancreas through the same midline incision. This pancreas-muscle flap is harvested two days later for assessment of viability.

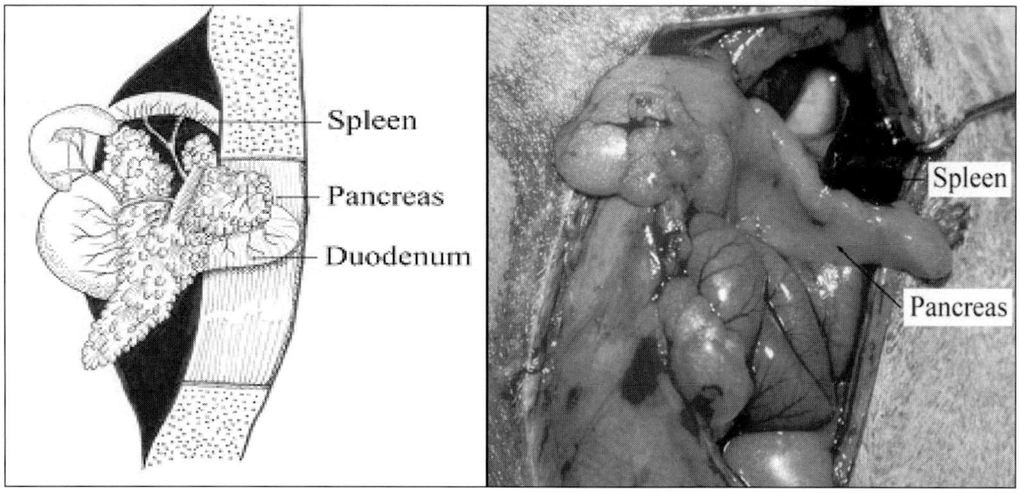

(C)

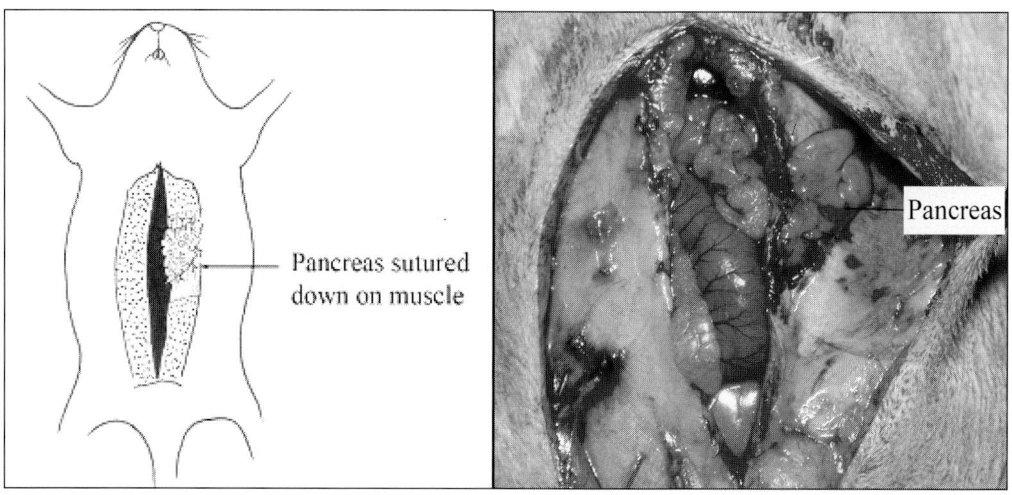

(D)

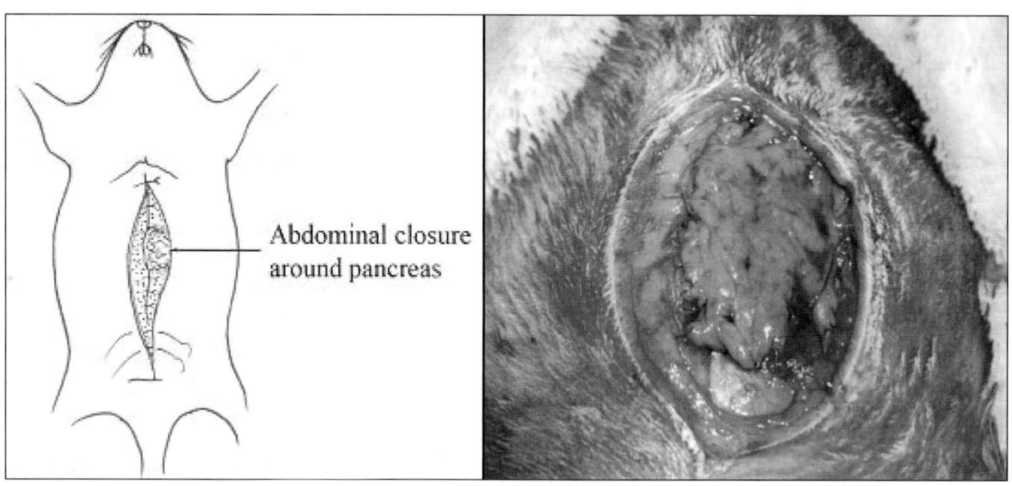

(E)

Fig 4.5.16 (***Continued***)

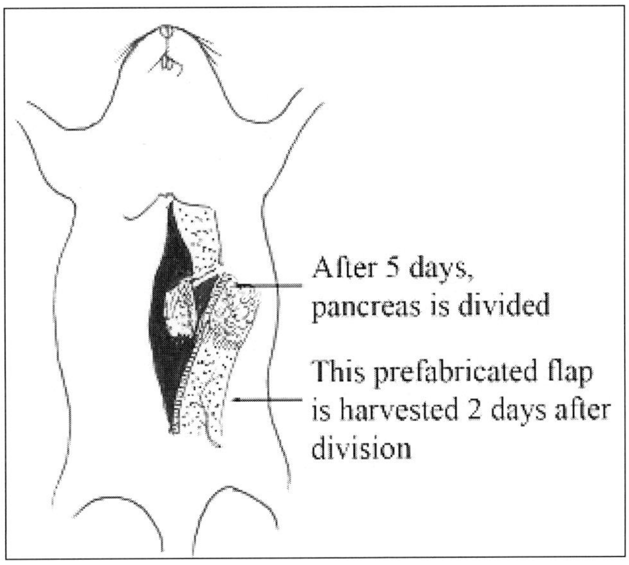

(F)

Fig 4.5.16 (*Continued*)

In the last phase, the viability of this neo-endocrine organ was demonstrated through histological examination (Fig 4.5.17). Histological examination was performed using H & E and immunochemical stains specific for insulin and vascular endothelium. More than 75 % of specimens showed viable islet cells that stained positive for insulin specific stains (Fig 4.5.18). The remaining 25 % showed poor uptake probably attributable to evolving surgical technique. This study proved that muscle flap vascularization of the pancreas was successful. Further preclinical studies in diabetic animal models would be necessary to evaluate its feasibility as a transplantable organ. This technique promises to be less invasive yet give long term insulin independence akin to whole organ transplants.

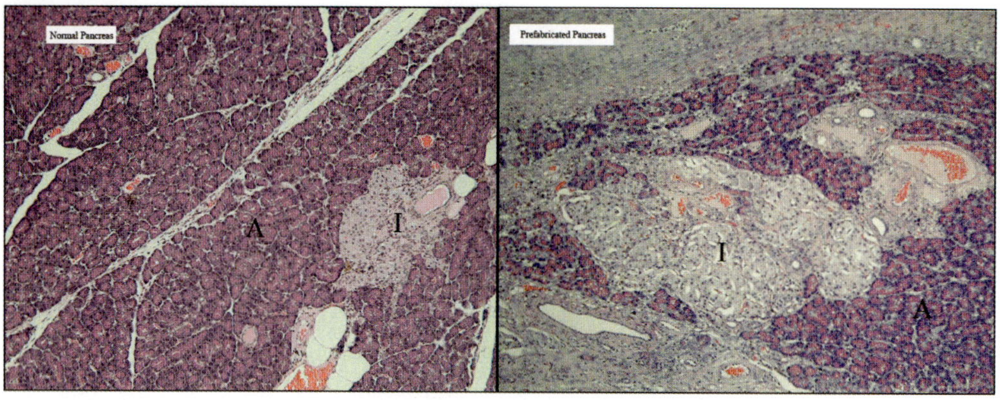

(A)

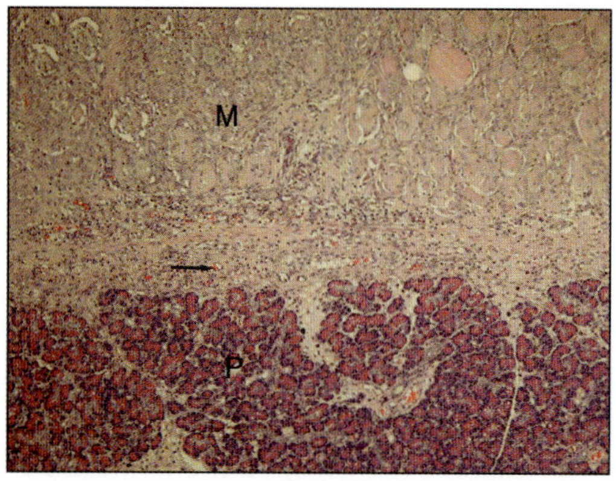

(B)

Fig 4.5.17: Histological examination of neo-pancreas.

(A) H & E at 100 ×. Normal pancreas (left) juxtaposed with neo-pancreas on muscle flap. Note that cellular architecture is preserved in the prefabricated pancreas and the similar appearance of the acinar (A) and islet (I) cells. (B) H & E at 200 ×. High power magnification of interface between the pancreas (P) and muscle (M) showing integration of the two tissues. The arrow shows a cross-section of a capillary traversing the muscle-pancreas interface.

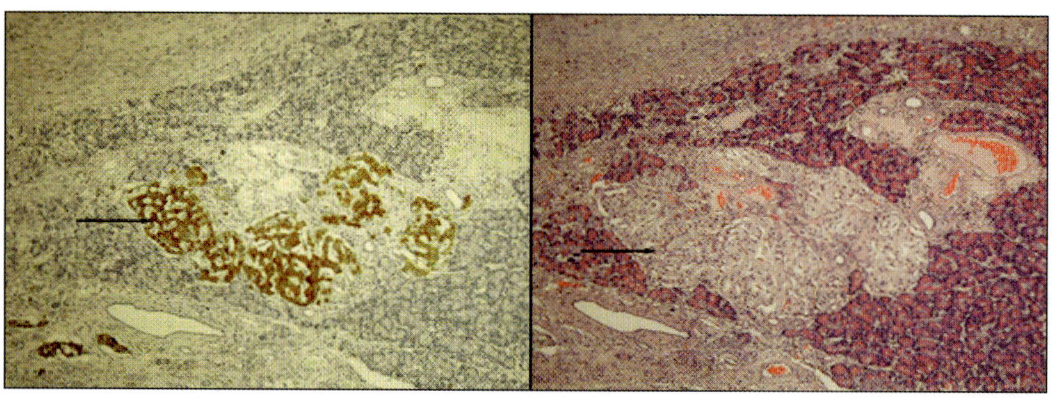

Fig 4.5.18: H & E and insulin immunostain at 100 × showing islet cells as indicated by arrow.
Note preservation of normal glandular architecture and positive insulin staining implying intact islet function.

Why is muscle such an effective vascularising agent? We postulate the following reasons:

- Muscle has stores of growth factors and angiogenic factors, which are liberated to promote tissue healing and adhesion (Frank J. M., 1998).
- Its extracellular matrix regulates angiogenesis and tissue repair in an orderly fashion. How this is achieved is unclear but the existence of regulation is evident. In the rat model, for example, we observed that viable prefabricated flaps did not exhibit florid neovascularization nor excessive scar proliferation in the bowel-muscle interphase as one would have expected. Instead, the junctions seemed quiescent, suggesting the presence of few but precisely connected vessels in a background of "minimal-scar" type tissue repair.
- Muscle has a dense capillary network of 2000 capillaries per square mm. By contrast, skin has 150 capillaries for the same unit area (Barker J.H. and Ryan T. J., 1998; Hill M. A. and Meininger G. A., 1998).
- Muscle has high blood flow — its resting blood flow is a fifth of cardiac output and four times that of skin. We postulate that this driving force contributes to angiogenesis (Hill M. A. and Meininger G. A., 1998).

Tissue Engineered Flaps

The ultimate goal of flap prefabrication is to generate vascularized tissues in proportions or compositions that are normally nonexistent. For example, large quantities of rib cartilage could be assembled to form an ear construct. This could, in turn, be transformed into a flap by inducing a suitable vascular carrier (for example, axial artery and vein, fascial flap) to perfuse it. One further step up the ladder of complexity would be to incorporate the concepts of tissue bioengineering for new tissue generation. Morrision *et al.* recently reported transforming adipocytes into mammary cells grown around arteriovenous bundles in tissue chambers. If these transformed cells remain differentiated and organized around the axial vessels, these tissues would qualify as tissue engineered flaps (Findlay M., Dolderer J., Cooper-White J. *et al.*, 2003; Dolderer J. H., Findlay M., Cooper-White J. *et al.*, 2003). By introducing alloplastic or bioengineered materials, flap composition could be further altered to create specialized tissue.

Tham and Song (2006) recently reported vascularising acellular dermal matrix (Integra®) with rat superficial inferior epigastric vessels and microsurgically transferring them as bioengineered dermal flaps. Flap prefabrication via vascular induction through staged transfer necessitates the harvest of a suitable vascular carrier, most commonly a piece of vascularized fascia. In this two-phase animal study, the authors explored the use of a bioengineered acellular dermal matrix (Integra®) as this carrier, avoiding the potential morbidities associated with harvest of a fasciovascular pedicle. Twenty young Wistar rats were used in each phase. In Phase I, twenty artificial dermal flaps were prefabricated by sandwiching the superficial inferior epigastric vessels of each rat within a 20 by 10 mm folded sheet of Integra®. The rats were subdivided into four groups of five rats each and were assessed at intervals ranging from 24 to 96 hours for signs of neovascularization of the dermal matrix. The earliest evidence of new vessel arborisation occurred within 24 hours and was detected on intravital microscopy and on histological examination. In Phase II, twenty flaps were fabricated as per phase one. After a lag period of 96 hours to allow the dermal matrix to become vascularised, free microsurgical transfer of the flap was performed (Figs 4.5.19 and 4.5.20).

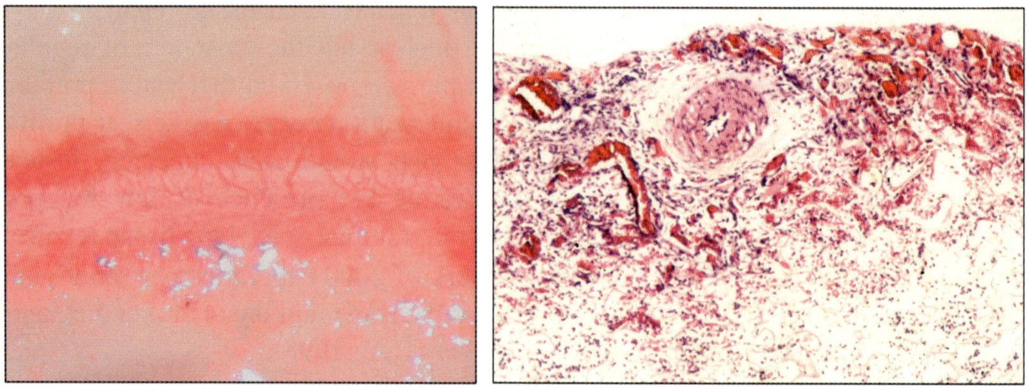

Fig 4.5.19: Ninety-six hours post fabrication.
(Left) Microscopic examination showing newly-formed vessels. *(Right)* Histological slide examination with routine staining. Notice the dense network of newly-formed vessels.

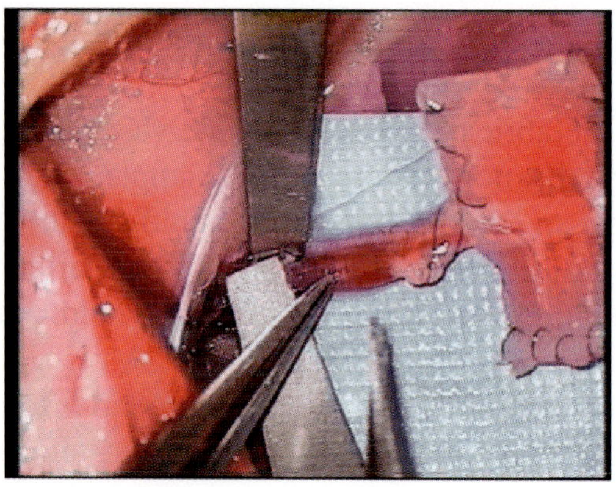

Fig 4.5.20: The Integra® flap undergoing microsurgical transfer using 11/O sutures under 20 × magnification.

Eighteen of the 20 flaps transferred were found to be viable 72 hours post-transfer. This was confirmed on histological examination, plain and intravital microscopy. The results showed that tissue-engineered matrices could be used as vascular carriers in free flap prefabrication. In addition, the potential for the fabrication of thin bioengineered dermal or skin flaps amenable to free microsurgical transfer, without major donor site morbidity may be possible (Tham C., Song C., Ng R. *et al.*, 2006).

CHAPTER 5

ANIMAL WELFARE CONSIDERATIONS

CHAPTER **5.1**

SPECIES SPECIFIC CAGING CONFIGURATION AND DESIGN

Cindy Phua

To minimize experimental variables, laboratory animals should be housed in caging systems that fulfil certain minimum requirements. These requirements can be classified into two categories, those that benefit the users and those that benefit the animals. The macroenvironment and facility design benefit both, while the microenvironment is of primary benefit to the animals.

Temperature, humidity, ventilation, light intensity and duration, noise and other variables contribute to the macroenvironment that directly affects the animal's cage environment, known as the microenvironment. The macroenvironment and microenvironment need to complement the facility design to ensure an optimum environment for the users and the animals.

There are three broad categories of facility design. They are barrier, containment and conventional facilities. A barrier facility is designed to keep out contamination whereas a containment facility designed to keep contamination from leaving the facility. Conventional facilities (often containing animals of mixed health status) may not have definitive design specification for "clean" to "dirty" traffic flow. Different types of animal will require different barrier needs, for example, specific pathogen free (SPF) animals may require a barrier facility for a more stringent barrier to prevent cross-contamination from conventional facilities, whereas animals subjected to infectious agents will need a containment facility to hinder the release of infectious contaminants.

Caging Systems

Caging system can be open or isolated. Open caging systems have a direct interaction with the macroenvironment. In open systems such as the metal rack caging and animal pens, animals are subjected to the variables of the room environment and users are also exposed to agents and allergens associated with the animals housed. These caging systems are common in barrier and conventional facility for large animals such as the rabbits and nonhuman primates.

Isolated caging systems are commonly used for small animals, like rodents, and when a high level of barrier and containment is required; examples of these are static micro-isolators and individually ventilated cages (IVCs) where an external fan unit ventilates the system. The fan unit supplies HEPA-filtered air to the cages at a high air exchange rate. While the older IVC systems only provide supply source, newer models have been developed to provide a controlled supply and exhaust function. This allows animal facilities to control the air pressure and the number of air changes in the cages. The air supply and exhaust can be adjusted for positive or negative pressure air balancing of the cage relative to the room. If set at negative, filters on the cage top keep contaminants out and thus, provide a clean environment for the animals. If the air balancing is set to make the cage positive to the room, the cage filter-top ensures that circulated air is filtered before it is released to the room and acts as a barrier for the cage if the fan unit fails. The filter-top also protects the animal when the cage is removed from the IVC rack. Exhausted air can be hard-ducted out of the room to reduce rodent odours. In high-level containment facilities, the exhausted air can be filtered before being released to the outside of the facility.

Cage Design

Cage design is dependent on the age, weight and size of the species. It should be escape-proof and made from caging material that is sturdy, durable and have a smooth and impervious surface for easy sanitation. The cage should also allow monitoring without disturbing the animal. The feeding tray and water source should be easily accessible by the animal and also be easily cleaned and disinfected. Fig 5.1.1 shows one such cages for rabbits.

Animal caging should provide sufficient space in accordance to the weight range of the species and the number of animal housed per cage. Animal facilities must comply with the recommended cage size listed in the "*Guidelines on Care and Use of Animals for Scientific Purposes*" by the National Advisory Committee for Laboratory Animal Research (NACLAR). Researchers who breach the guidelines will be reported to the Institutional Animal Care and Use Committees (IACUC) and may have their projects suspended.

Due to its strength, durability and the ability to withstand repeated sanitation, stainless steel is a common material used. This material is more cost-efficient as the need to replace cages will be reduced. The strength of stainless steel makes it suitable for housing large animal due to their weight. Moreover, stainless steel does not corrode with most disinfectants and can withstand repeated high temperature washing.

Open rack caging can be made fully or partially from metal, usually stainless steel. They consist of perforated flooring that allows waste to be collected into a tray. Open rack caging can be modified to suit the user's and the animal's needs. For example, nonhuman primate cages (Fig 5.1.2) are equipped with a perch and an additional back panel that can be moved, called the squeeze-back. The squeeze-back mechanism moves the animal to the front of the cage to partially immobilize the animal for manipulation through the cage front or for injections. Another example of the open system is the animal pens, commonly used to house larger animals such as swine and sheep. Animal pens consist of a stainless steel framework and a non-slip impervious flooring that is easily disinfected and usually equipped with a good drainage system.

While IVC racks are generally made from stainless steel, the individual cages are usually made from plastic polymers. The types of plastic polymers range from the most basic polystyrene to thermosplastic such as polysulfone and polyetherimide. Polystyrene cages cannot endure repeated high temperature washing and autoclaving but can be disinfected. They are frequently used in research where routine decontamination is not feasible, such as research where radioactive and hazardous agents are used. Polycarbonate cages are commonly used, but the lifespan of use is shortened by frequent autoclaving. Polysulfone and polyetherimide cages have a longer lifespan and can withstand high temperatures of 150°C and 160°C respectively. These thermoplastic cages can be autoclaved to house SPF and immunodeficient animals.

Fig 5.1.1: Rabbit cage.

Fig 5.1.2: Meshed cage – monkey.

Cage Accessories

Cage enrichments are required for almost all species of animals especially social animals and individually housed animals. These provide stimuli that encourage the expression of species-typical behaviour. Individual housing can be traumatic to the animals and enrichment can reduce this stress.

Enrichment accessories such as disposable paper roll and polymer vinyl chloride (PVC) cylinder (Fig 5.1.3) provide shelter and gnawing for the rodents. Running wheel also provide enrichment for rodents and may be encouraged for breeding colonies where overweight breeders are discouraged.

Enrichment toys such as a plastic chain or plastic ball are commonly used for most large animals for gnawing and visual stimuli, such as those for pigs (Figs 5.1.4 and 5.1.5) and rabbits (Fig 5.1.6). Heavy plastic is often used as it is durable, economical and can be easily disinfected.

Nonhuman primates require a more intensive enrichment programme due to their complex needs. Group or pair housing should be provided whenever possible for social enrichment. Visual, auditory or olfactory contact with other primates should be provided if animals must be individually housed. In some cases, grooming contact can be provided while keeping animals in individual cages. Nonhuman primates should be provided with a perch, manipulanda such as toys and foraging devices or opportunities. Over-grooming commonly associated with individually housed animals can sometimes be reduced by providing complex, time consuming foraging opportunities. Mirrors or reflective toys will also help provide visual enrichment. Fruit and/or vegetable treats can also be a part of an enrichment programme, but should not be allowed to replace the balance ration. If using outdoor pens for nonhuman primates the weather should be taken into consideration. Rain cover, shade and protection from temperature extremes should also be provided.

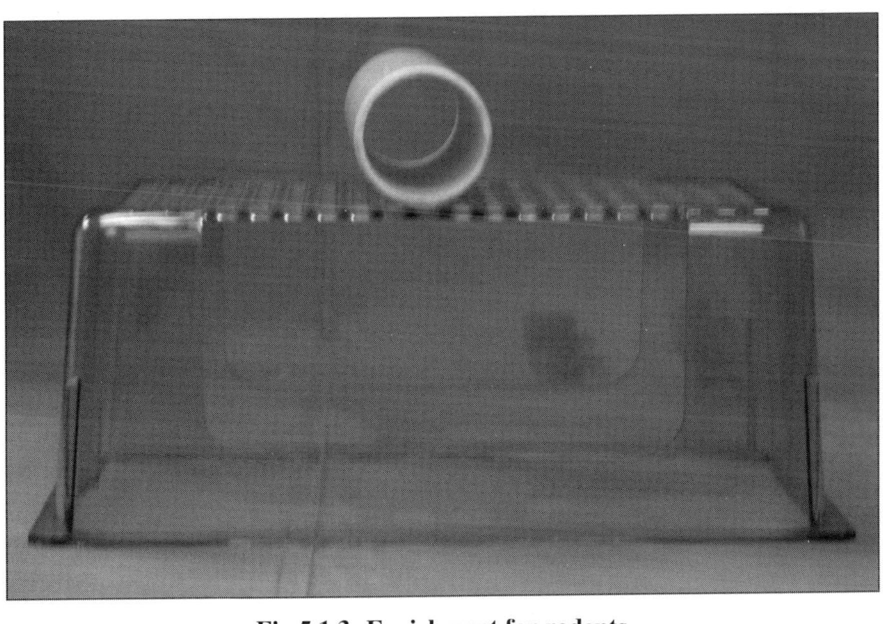

Fig 5.1.3: Enrichment for rodents.

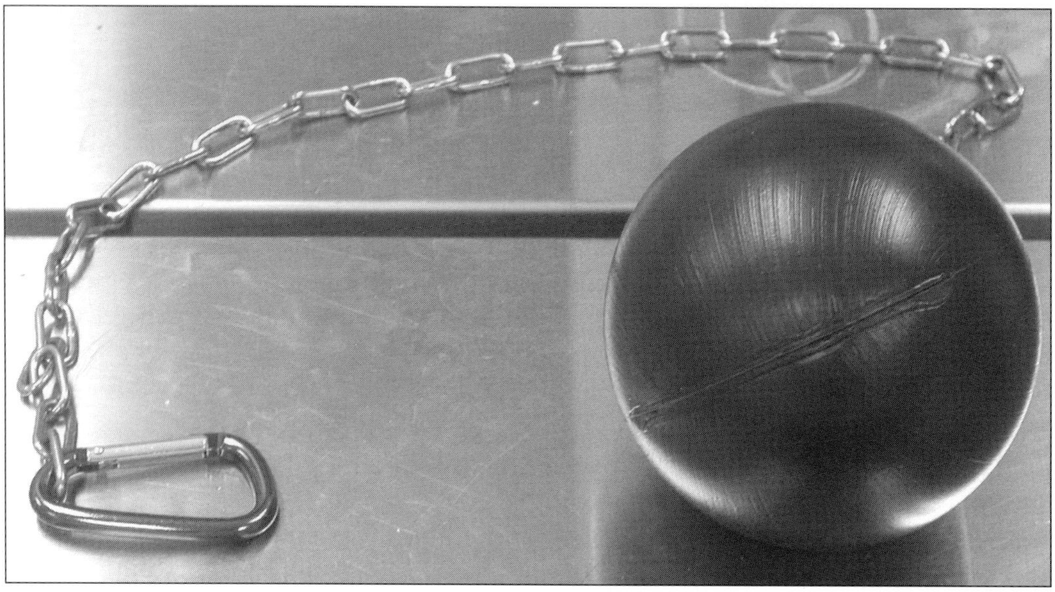

Fig 5.1.4: Enrichment I for pigs.

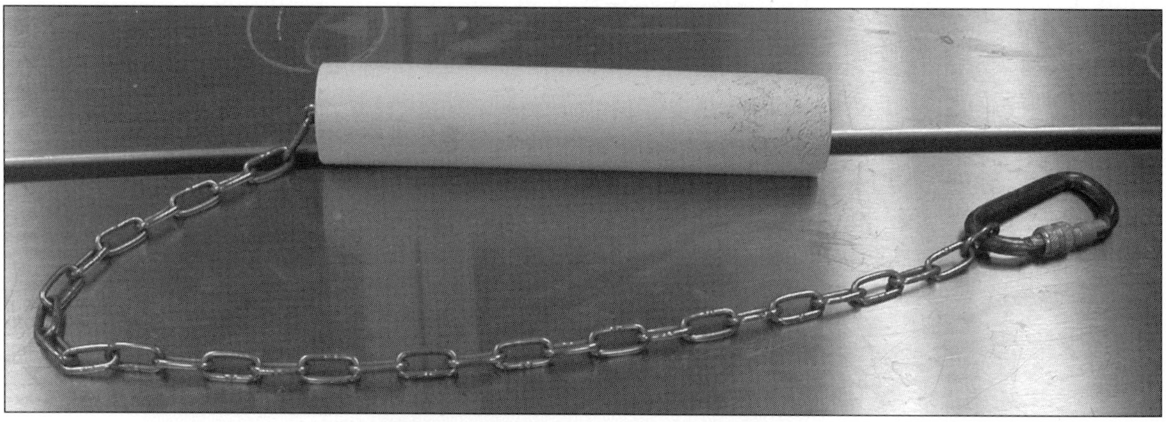

Fig 5.1.5: Enrichment II for pigs.

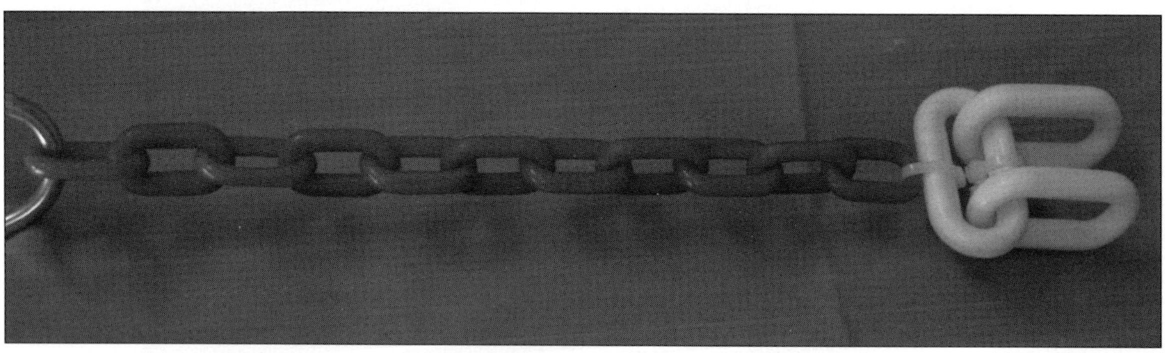

Fig 5.1.6: Enrichment for rabbits.

Other Caging Systems

1. *Transport Cages*
   Transport cages provide mobility and temporary housing for animals. These cages should have adequate room for the animals, be escape-proof and well ventilated. While transport cages provide temporary housing, for example when the animal cages need to be sanitized, the animals should not be housed in them for prolonged periods.

2. *Recovery Cages*
   A recovery cages or intensive care unit (Fig 5.1.7) are used in post surgical recovery and for clinical care of sick animals. They are usually mobile and constructed of fibreglass. The units are also fitted with environmental controls that regulate humidity, temperature and oxygen levels.

Fig 5.1.7: Intensive care unit/recovery cage.

3. *Restrainers*
   Restraining devices are designed to immobilize animals for short-term handling while they are being treated or manipulated. Specific restraint device depends on the species being restraint. Plastic rodent cyclinder (Fig 5.1.8), rabbit restraint box (Fig 5.1.9), pig slings and nonhuman primate chairs (Fig 5.1.10) are examples of restrainers.

Fig 5.1.8: Rodent restrainer.

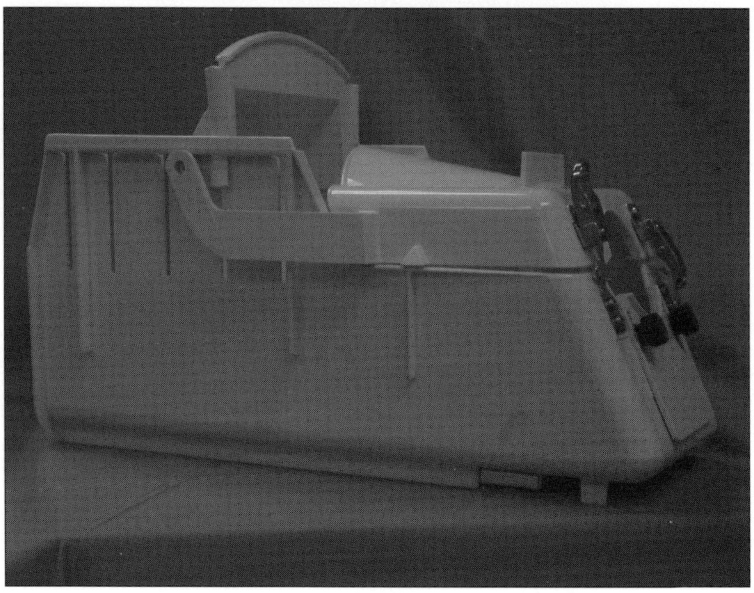

Fig 5.1.9: Rabbit restrainer.

Fig 5.1.10: Nonhuman primate restrainer.

Standards for Housing and Environmental Conditions

Tables 5.1.1, 5.1.2 and 5.1.3 list the standards practice for housing of different types of animals and environmental conditions.

Table 5.1.1: Goat, sheep and swine

Animal	Weight/kg	Floor area/m^2	Height/cm
Goat/Sheep	< 25	0.8/0.765/0.675	-
	Up to 50	1.35/1.125/1.017	
	> 50	1.8/1.53/1.35	
Swine	< 15	0.72/ - / -	-
	Up to 25	1.08/0.54/0.54	
	Up to 50	1.35/0.9/0.81	
	Up to 100	2.16/1.8/1.62	
	Up to 200	4.32/3.6/3.24	
	> 200	≥ 5.4/≥ 4.68/≥ 4.32	

Table 5.1.2: Rabbit, cat, dog, chicken, nonhuman primate

Animal	Weight/kg	Floor Area/m^2	Height/cm
Rabbit	< 2	0.135	35
	Up to 4	0.27	
	Up to 5.4	0.36	
	> 5.4	> 0.45	
Cat	≤ 4	0.27	60
	> 4	≥ 0.36	
Dog	< 15	0.72	-
	Up to 30	1.08	
	> 30	≥ 2.16	
Chicken	< 0.25	0.225	-
	Up to 0.5	0.045	
	Up to 1.5	0.09	
	Up to 3.0	0.18	
Monkey	Up to 1	0.144	50
	Up to 3	0.27	76
	Up to 10	0.387	

Table 5.1.3: Mouse, rat, hamster and guinea pigs

Animal	Weight/kg	Floor Area/m^2	Height/cm
Mouse	< 10	38	12
	Up to 15	51	
	Up to 25	77	
	> 25	> 96	
Rat	< 100	109	17
	Up to 200	148	
	Up to 300	187	
	Up to 400	258	
	Up to 500	387	
	> 500	< 451	
Hamster	< 60	64	15
	Up to 80	83	
	Up to 100	103	
	> 100	> 122	
Guinea Pig	≤ 350	387	17
	> 350	> 651	

* Larger animals might require more space to meet performance standards

CHAPTER 5.2

POSTOPERATIVE CARE AND PAIN MANAGEMENT

Jason Villano

Any manipulation of laboratory animals such as handling and physical restraint, surgical procedures and even routine procedures such as blood collection can have profound effects on their behaviour and physiology, which can variably be reflected in the research results. Inasmuch as unnecessary data or data misinterpretation are concerns in the quality of research derived from these animals, animal welfare posts a significant role in the conduct of these experiments, particularly those involving operative procedures.

The research personnel led by the principal investigator and the animal care staff led by the institutional veterinarian are the two groups that play key roles in developing an effectively managed postoperative care programme tailored to the institution's needs. The veterinarian and his staff can recommend and implement measures on alleviating the pain and distress of the animals while it is the researcher's responsibility to show that he has considered all these options without compromising the validity of the research. Effective communication between these two groups is thus essential.

Postoperative Support

The postoperative period can be divided into three phases, anaesthetic recovery, acute postoperative care, and long-term postoperative care.

1. *Anaesthetic Recovery*
   Frequent and careful observation is required during this stage as it is the most critical. Great physiologic disturbance and crises can arise quite rapidly at this time. Large

animals, whose trachea is intubated for inhalation anaesthesia during the surgery, can vomit and suffer from aspiration pneumonia. The animal should only be extubated when the gagging or swallowing reflexes have returned. The animal's vital signs, cardiovascular and respiratory functions must also be checked and maintained. Rotating or turning over the animal's body every 30 to 60 minutes until it has recovered from the anaesthesia will facilitate respiration and avoid dependent oedema. Usually, animals should be individually housed during recovery in cages that have been sanitized between usage.

2. *Acute Postoperative Care*
During this stage, the animal is usually maintained in the recovery area such as an ICU (Fig 5.2.1) until adequate stabilization allows removal to a more standard husbandry situation (that is, eating and drinking have resumed and critical physiological parameters are within acceptable ranges for the model created). Pain management should be started at this time or continued if preemptive analgesia has been given. The investigator and/or the veterinary staff must be familiar with the animal's normal behaviour and posture considering species and individual variation. Unless there is evidence to the contrary, it has to be assumed that a procedure or a condition painful for humans will also be painful for animals.

Besides the analgesics, parenteral fluids and antibiotics may be continued. The hydration status should be monitored as overhydration results in frequent urination and pulmonary oedema while underhydration results in sticky mucous membranes, loss of skin elasticity, the eyes sinking into the orbit, decrease in blood pressure and increase in heart rate. Replacement of blood loss is with saline or Lactated Ringers Solution administered three times the volume of blood lost by slow intravenous drip. Monitor the haematocrit. If it drops below 20 %, whole blood replacement may be necessary.

Fig 5.2.1: Intensive care unit.

3. *Long-Term Postoperative Care*
   Although it is often the most neglected phase in the clinical monitoring, long-term management is equally important as it returns the animal to as normal a physiological and behavioural state as possible. This includes wound management and monitoring of appetite, body weight and activity.

Wound management prevents infection and inflammation and facilitates the healing process. When drains, collars and dressings are used, the animal's ability to eat and drink should not be hampered. If the wound is exposed, daily cleaning and monitoring needs to be done to remove accumulated dirt such as faeces. Chlorhexidine or iodine swabs and antibiotic ointments or powder may be used (Fig 5.2.2). Conversely, if the wound is covered with a dressing, regular cleaning and changing should be done as often as every other day or as necessary when the dressing is wet. Any external sutures are removed once skin incision site is healed, usually between 10 to 14 days.

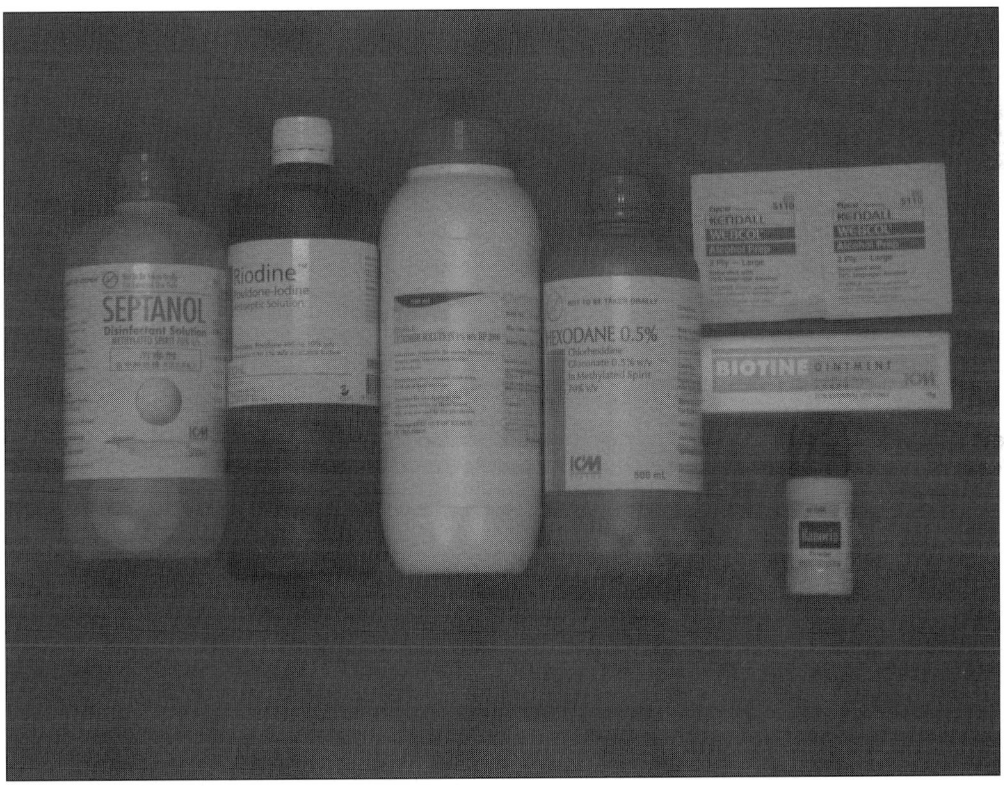

Fig 5.2.2: Commonly used disinfectant and topical antibiotics.

The quantity and quality of the faeces and urine should also be monitored because changes may indicate several postoperative complications such as paralytic ileus, renal shutdown or irritation hypermotility. Regular checks on the body weight and appearance and the animal's appetite should also be done. Physical therapy may also be needed in some cases for postoperative paresis or paralysis.

Pain Management

This is an important aspect in the perioperative care as it has profound effects on the animal's physiology and behaviour and it addresses the issues and concerns of animal welfare (Fig 5.2.3). An animal's response to pain is often adaptive to reduce movement minimizing re-injury and aiding recuperation. However, this response may lead to changes which impact negatively on both the animal's well-being and research results.

Examples of procedures that may cause pain or distress are physical restraints, survival surgeries, tumour burdens, intracardiac or orbital sinus blood sampling, and abnormal environmental conditions. These procedures can cause changes in the heart rate, blood pressure, respiration and body temperature. Blood glucocorticoid and catecholamine levels are also usually elevated.

Otherwise, it is difficult to assess pain and distress in animals because of their inability to communicate directly. Because of this, animal welfare regulations require that analgesia be provided whenever a procedure is to be performed or a condition is present that is likely to cause pain. It is best if analgesia can be provided to animals preemptively.

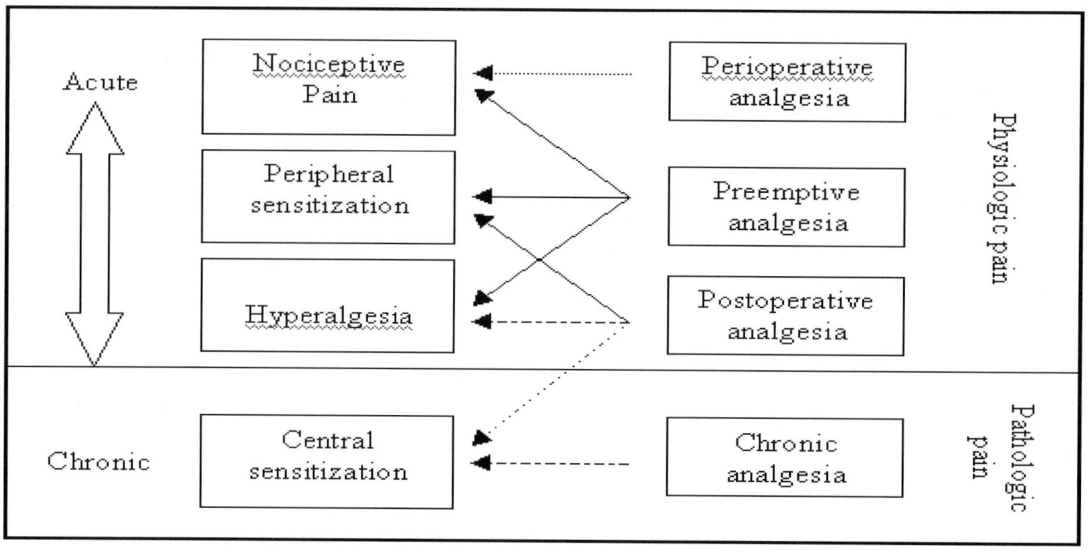

Fig 5.2.3: Analgesic strategies for pain management. The temporal progression of pain and approaches (and their targets) to the initiation and maintenance of analgesia. Arrows show effector pathways (dotted lines indicate lower efficacy). Adapted from Kissin I, Preemptive analgesia, *Anesthesiology* 2000; 93: 1138–1143.

The following is a guide to the species variability of animal in response to pain. It is important to note that individual variation also exists and may be especially pronounced in higher forms of animals like the nonhuman primates.

- Nonhuman primates — hunched posture, failure to groom, refusal of food or water, dejected appearance.
- Mice — withdrawal, biting response, piloerection, hunched posture, sunken eyes and abdomen, dehydration, weight loss.
- Rats — vocalization, struggling, licking/guarding, weight loss, piloerection, hunched posture, hypothermia.

- Rabbits — reduced eating and drinking, facing towards back of cage, limited movement, apparent photosensitivity.
- Pigs — vocalization and/or the lack of normal social behaviour, reluctance to move.
- Sheep and goats — rigid posture and reluctance to move.

Analgesic agents (Table 5.2.1) include non-steroidal anti-inflammatory drugs (NSAIDs), glucocorticoids and narcotics. NSAIDs belong to a group of drugs having analgesic, antipyretic and anti-imflammatory activity due to their ability to inhibit the synthesis of prostaglandins. It includes aspirin, paracetamol, phenylbutazone, carprofen and ibuprofen. Meanwhile, glucocorticoids such as dexamethasone are also used for inflammation but care should be taken especially when the animal is pregnant as these can terminate the pregnancy in some species. Narcotics such as opiod derivatives morphine and etorphine are more potent than NSAIDs and glucocorticoids, but they can cause sedation, reduced GI motility, respiratory depression and they lack anti-inflammatory activity. Fentanyl transdermal patch can also be used for chronic management of pain in certain animals like dogs, pigs and rabbits. They are not generally used in nonhuman primates (NHPs) due to difficulty in preventing removal by the dexterous hands of the animals. Use in sheep is also problematic since the natural lanolin secretion from their skin reduces the adherence of the patch, reducing the potential for transdermal absorption.

Table 5.2.1: List of drugs commonly used in the laboratory animal

Drug	Mice	Rats	Rabbit	Pigs	Sheep	Nonhuman primate	Hamsters
Buprenorphine	0.05-0.1 mg/kg SC, IV q8-12h	0.1-0.5 mg/kg SC, IV q12h	0.02-0.05 mg.kg SC, IM, IV q8-12h	0.05-0.1 mg.kg IM, SC q8-12h	0.005 mg/kg SC, IM q8-12h	0.01-0.03 mg/kg IM, SC q8-12h	0.05-0.1 mg/kg, SC, IM q8-12 h
Butorphanol	1-5 mg/kg SC q6-8h	0.05-2.0 mg/kg SC q4h	0.1-0.5 mg/kg IV q4h	0.1-0.3 mg/kg SC, IM, IV q6-8h	0.3 mg/kg SC, IM, IV q12h	0.025 mg/kg IM q3-6h	1-5 mg/kg SC, Im q2-4h
Flunixin meglumine	2.5 mg/kg SC, IM try q12h	1.1 mg/kg SC, IM q12h	1.1 mg/kg SC, IM try q12h		0.5-1.0 mg/kg IV, IM q8h	0.5 mg/kg IM q24h	2-5 mg/kg, SC, IM q12-24h
Morphine	2.5 mg/kg SC q6-8h	10 mg/kg SC q2-4h	2-5 mg/kg SC IM q2-4h	0.2-0.9 mg/kg SC	-------	1-2 mg/kg IM, SC q4h	2-5 mg/kg, SC, IM q2-4h
Acetaminophen	300 mg/kg PO	110-300 mg/kg PO	1ml elixir in 100ml drinking water	-------	NR	10 mg/kg PO q8h	-------
Asprin	20 mg/kg SC, 100-120 mg/kg PO	20 mg/kg SC, 100-120 mg/kg PO	100 mg/kg PO q12h	10-20 mg/kg PO q8h	-------	10-20 mg/kg PO q6h	100-150 mg/kg PO q4h
Carprofen	5 mg/kg SC q24h	5 mg/kg SC q12-24h	1.5 mg/kg PO q12h	0.5-4.0 mg/kg SC q24h	4 mg/kg SC q72h	2 mg/kg PO SC q12-24h	5 mg/kg, SC, PO q12h
Ketoprofen	5 mg/kg PO, SC q24h	5 mg/kg PO, SC q24h	1 mg/kg IM	1.0-3.0 mg/kg SC, IM q24h	3 mg/kg IV, IM q24h	2 mg/kg IV IM q24h	2 mg/kg, SC q2-4h

Abbreviations: SC – subcutaneously, IV – intravenously, IM – intramuscularly, PO – orally, NR – not recommended.

Aside from management of pain using chemicals, cold compresses can also be intermittently applied on the region of interest in the first 24 to 48 hours. Application should just be long enough to produce vasoconstriction and can be as short as 20 seconds. Environmental considerations like placing cushions on the flooring to alleviate the pressure and prevent pressure sores and adding enrichment devices can also be done.

CHAPTER 5.3

ANIMAL FEEDS AND NUTRITIONAL REQUIREMENTS

Peik Khin Tan

Adequate nutrition for laboratory animals consists of water and food containing nutrients essential to provide energy and raw materials for growth, maintenance and repair of body tissue. Food items are composed of water, proteins, fats, carbohydrates, vitamins and minerals. Each nutrient type plays specific roles for different body processes are essential to animal health and well-being.

Nutritional requirements vary depending on species (e.g., herbivores, carnivores, omnivores), stages of life (e.g., reproduction, growth, maintenance), health status or condition (e.g., allergies, urinary tract infection), research protocol (e.g., *ad libitum*, restricted amount), gender and environmental condition (e.g., temperature, humidity).

Most laboratory animal diets contain 18 to 25 % crude protein of animal or plant origin, providing all essential amino acids in the right proportions. The largest percentage of total diet is usually carbohydrates, usually vegetable origins, mainly cereals, which serves as a source of energy. Most diets are composed of 2 to 8 % fat from plant or animal origin, which serves as a vehicle for fat-soluble vitamins and provides essential fatty acids. Fibre contents in all diets are made from natural ingredients. Water-soluble vitamins are usually found in the non-fatty tissues of plants and animals while fat-soluble vitamins can be obtained from the fatty parts and oils in plants. Minerals such as calcium and phosphorus are required for teeth and bone growth. Sodium and potassium play an important role in acid base balance. Any deficiency or excess in vitamins and minerals can cause serious disease.

Animals are provided with clean feed and water free of pathogenic organisms and harmful chemicals. The commercially available animals' feeds must meet the nutritional requirements of each species. Feed and water supplied to animals are tested quarterly to ensure they are safe and the nutrients values are within the ranges stated on the label of the feed bag.

Rodents, rabbits, pigs, sheep and geese are fed pelleted diets while tree shrew and monkeys are fed monkey chow formed into biscuits. Specific Pathogen Free (SPF) rodents are given autoclaved food and water. Caution must be taken though when using autoclaved feeds as the moisture during the autoclaving process can facilitate fungus build-up, vitamins can be destroyed and proteins can caramalize, making the pellets hard. Irradiated feeds (gamma radiation) are also available commercially.

The daily feed and water requirements for laboratory animals are listed in Table 5.3.1.

Table 5.3.1: Daily feed and water requirements for selected laboratory animals

Species	**Daily feed intake (gms)/100 g BW/day**	**Daily water intake (ml)/100 g BW/day**
Mouse	12–15	15
Lactating Mouse	80–100	80–100
Hamster	5	10
Rat	5	10
Rabbit	5	10
Lactating Rabbit	10–15	Up to 90
Monkey (g/kg BW/d)	350–550	350–950
Domestic Swine (kg/d)	3.6–4.1	80–120
Sheep (g/kg/d)	15–60 (dry matter intake)	197

Nutritional Requirements for Rodents

Rodents (mice and rats), being omnivorous, are fed with 12 mm diameter rat and mouse pellets fortified with vitamins and minerals to meet the requirements of these animals after the diet is autoclaved (Fig 5.3.1). The diet is made of wheat, lupins, barley, soya meal, fish meal, mixed vegetable oils, canola oil, salt, calcium carbonate, dicalcium phosphate, magnesium oxide, vitamin and trace mineral premix to contain 19.6 % protein, 4.6 % total fat, 4.5 % crude fiber and 14.3 mj/kg digestible energy. This commercially available rat feed is also generally used as the basic diet for hamsters, sometimes in combination with rabbit food to provide a balance of 16 to 24 % protein, 60 to 65 % carbohydrate and 5 to 7 % fat.

Fig 5.3.1: Rodent pellet.

The added vitamins and trace minerals are as follows (Table 5.3.2):

Table 5.3.2: List of added vitamins and minerals in mouse and rat diet

Added vitamins and minerals in the mouse and rat diet	
Vitamin A (Retinol)	10,000 IU/Kg
Vitamin D3 (Cholecalciferol)	2000 IU/Kg
Vitamin K (Menadione)	20 mg/kg
Vitamin E (α Tocopherol acetate)	100 mg/kg
Vitamin B1 (Thiamine)	80 mg/kg
Vitamin B2	30 mg/kg
Niacin (Nicotinic acid)	100 mg/kg
Vitamin B6 (Pyridoxine)	25 mg/kg
Calcium Pantothenate	50 mg/kg
Biotin	300 mg/kg
Folic acid	5 mg/kg
Vitamin B12 (Cyanocobalamin)	150 µg/Kg
Magnesium	100 mg/kg
Iron	70 mg/kg
Copper	16 mg/kg
Iodine	0.5 mg/kg
Manganese	70 mg/kg
Zinc	60 mg/kg
Molybdenum	0.5 mg/kg
Selenium	0.1 mg/kg

Nutritional Requirements for Rabbits

Rabbits are herbivorous and given 4-mm diameter, 4 to 10 mm long meat-free guinea pig and rabbit pellets and timothy hay twice a day. The diet (Fig 5.3.2) is fortified with vitamins and minerals to meet the daily nutritional requirement. The diet is composed of lupins, oaten hay, Lucerne, barley, soya meal, canola meat, DL Methionine, mixed vegetable oils, salt, dicalcium phosphate, magnesium oxide, a vitamin and trace mineral premix to make up 18.2 % protein, 4.2 % total fat, 14.4 % crude, 18 % acid detergent fibres and 11 mj/kg digestible energy.

Fig 5.3.2: Rabbit feed.

The added amino acids, vitamins and trace minerals are as listed in Table 5.3.3:

Table 5.3.3: List of added vitamins and minerals in rabbit diet

Added vitamins and minerals in the rabbit diet	
Vitamin A (Retinol)	49, 000 IU/kg
Vitamin K (Menadione)	3 mg/kg
Vitamin E (α Tocopherol acetate)	60 mg/kg
Vitamin B1 (Thiamine)	5.6 mg/kg
Vitamin B2	6.6 mg/kg
Niacin (Nicotinic cid)	56 mg/kg
Vitamin B6 (Pyridoxine)	5.8 mg/kg
Pantothenic acid	19 mg/kg
Biotin	140 μg/kg
Folic acid	0.6 mg/kg
Vitamin B12 (Cyanocobalamin)	7 mg/kg
Choline	2,200 mg/kg
Iron	40 mg/kg
Copper	13 mg/kg
Iodine	1.7 mg/kg
Manganese	90 mg/kg
Cobalt	0.7 mg/kg
Zinc	60 mg/kg
Selenium	0.1 mg/kg

Nutritional Requirements for Nonhuman Primates (NHPs)

All nonhuman primates (NHPs) require Vitamin C in their diet like man and guinea pigs. These species of animals cannot synthesize vitamin C unlike most because they lack the enzymes necessary for conversion of L-gluconolactone to L-ascorbic acid and cannot store the vitamin in any appreciable extent. Lack of Vitamin C leads to scurvy, which causes formation of livid spots on the skin, spongy gums, and bleeding from almost all mucous membranes.

Fig 5.3.3: Monkey Chow.

Nutritional requirements vary from Old World Monkey to New World monkey. *Macaca fascicularis* (cynomolgus or crab-eating macaque) is under Old World Monkey species and fed with Laboratory Fiber-Plus® Monkey Diet (Fig 5.3.3). The diet is made of ground corn, dehulled soybean meal, ground soybean hulls, ground oats, corn gluten meal, ground wheat, animal fat preserved with BHA, dehydrated alfalfa meal, sucrose, dicalcium phosphate, dried whey, fish meal, calcium carbonate, brewers dried yeast, salt, L-ascorbyl-2-polyphosphate, pyridoxine hydrochloride, menadione dimethylpyrimidinol bisulfite, cholecalciferol, DL-methionine, choline chloride, vitamin A acetate, folic acid, calcium pantothenate, ferrous sulfate, dl-alpha tocopheryl acetate, biotin, thiamin mononitrate, nicotinic acid, riboflavin, cyanocobalamin, zinc oxide, L-lysine, manganese oxide, ferrous carbonate, copper sulfate, zinc sulfate, calcium iodate, cobalt carbonate, sodium selenite. This diet contains 20 % crude protein, 5 % crude fat, 10 % crude fibre, 3 % added minerals and 5.5 % ash.

Monkey Chow must be used within 180 days of manufacture, assuming it contains a stabilized form of Vitamin C, otherwise it must be used in 90 days. The stability of Vitamin C varies with environmental conditions, therefore special care must be taken to store feed properly. Monkeys generally consumes about 2 to 4 % of their body weight in food each day. The daily food allowance is given in equal portions twice during the day to prevent wastage. Fresh, clean water is available at all times from an automatic watering system. Laboratory Fiber-Plus® Monkey Diet is sometimes soaked in fruit juice to soften the product for infants or animals that have difficulty chewing.

Fruits such as bananas, apples, grapes and oranges are given once a day for primate enrichment, assuming it does not interfere with the research protocol.

Nutritional Requirements for Swines

Gold Coin Company from Malaysia supplies swine feed (Fig 5.3.4). The feed is made of yellow maize, molasses sugarcane, canola meal, high protein soyabean meal, palm oil, rice bran, wheat pollard, L-lysine, methionine, L-threonine, dicalcium phosphate, limestone dust, salt, sodium bicarbonate pig mineral mix, choline chloride liquid, and pig vitamin basemix. The diet contains 17.5 to 19.5 crude protein, 3.5 to 7 % ether extract, 2.5 to 5 % crude fibre, 0.9 to 1.1 % calcium, 0.6 to 0.8 % phosphorus and 13 % maximum moisture content. The diet is fortified with minimum values of 7 IU vitamin A, 1.2 IU vitamin D, 10 g vitamin E, 1 g vitamin K, 100 g zinc, 150 g copper and 100 g ferrous per ton of feed. Swine consume 1 to 4 % of body weight. Swine do not require elemental sulfur since they utilize sulfur-containing amino acids. The water is supplied by automatic watering system *ad libitum*.

Fig 5.3.4: Pig pellet.

Nutritional Requirements for Other Laboratory Animals

The nutritional requirements of tree shrews are not well documented at this moment although they are fed once daily with LabDiet® monkey chow. Fruit supplements like bananas, apples and pears are provided as well. Each tree shrew is to be provided with 14 g of food (2 monkey chows) per day. Commercial dog food is also found to be suitable for these animals.

Geese are fed once daily with Gold Coin Pig Grower Feeds unless otherwise instructed. Feeds like grass are provided as often as possible. On zero grazing, geese will eat up to 200 g of food per day (depending on the size of the goose).

Sheep feed is also supplied by Gold Coin and it contains minimum crude protein 15 to 17 %, maximum crude fibre 12 %, minimum crude fat 3 %, maximum moisture 13 %, maximum ash 12 %, calcium 0.8 to 1.4 %, and phosphorus 0.5 to 0.9 %. The sheep feed is made of molasses, wheat, soyabean meal, palm oil, rice bran, cocoa cake, wheat pollard, sodium bicarbonate mineral, choline chloride, and vitamin basemix. Sheep are also given high quality hay from Australia. The hay provides 14.2 % protein, 7.4 % ash and minerals.

It is very important to note that sheep is unique among food and farm animals in the way they utilize copper. Copper, a required mineral, is potentially toxic to all food animals although sheep is most susceptible. Its metabolism is affected by the presence of other minerals and some ionophores, especially the levels of molybdenum and sulfur, which act as its antagonists. These compounds bind with copper and prevent gut absorption and increase excretion of absorbed copper in the liver and body tissues. Prevention of copper toxicity involved not feeding sheep any swine, cattle or poultry rations, which contain high levels of copper by design.

The feed samples are analyzed once a year to ensure the nutrient composition mentioned by manufacturers contains in the feed.

CHAPTER 6

SAFETY MANAGEMENT OF AN ANIMAL FACILITY

CHAPTER **6.1**

OCCUPATIONAL HEALTH AND SAFETY PROGRAMME

Angela Goh

In recent years, there has been an increase in attention on biosafety and occupational health in Singapore. Reflecting this trend, the Singapore General Hospital (SGH) has moved towards accreditation of its healthcare services, and has been awarded the ISO 14000 (Environmental Management Standard), ISO 18000 (Occupational Health and Safety Assessment), and Joint Commission International (JCI) certifications. Being a part of SGH, the Department of Experimental Surgery (DES) is directly involved in the accreditation process and is committed to better organization, improved animal care and a safer working environment with reduction of risk to staff and researchers.

On 15th October 2002, the SGH-IACUC (Institutional Animal Care and Use Committee) was formed as a pre-requisite to DES effort to seek accreditation from Association for the Assessment and Accreditation of Laboratory Animal Care (AAALAC). An important component of the AAALAC guidelines is the incorporation of an Occupational Health and Safety programme in assuring a safe environment for the conduct of animal research activities. In addition, the establishment of the SGH-EHS (Environmental Health and Safety) Committee on 1st April 2004 provided for a comprehensive coverage of environmental control and biosafety issues in the hospital. DES proposed the establishment of two committees to oversee EHS implementation particularly for animal facilities. In March 2005, the Animal Facilities Biosafety Committee and the Committee for Emergency Crisis Management for Animal Facilities were formed. Both these committees assist the department in overseeing biosafety, occupational health, and crisis management issues.

AAALAC requirements for an occupational health and safety programme for personnel working with laboratory animals are detailed in the publication *Occupational Health and Safety in the Care and Use of Research Animals* (published by the U.S. National Academy of Sciences). Following these guidelines, DES established an occupational health programme in 2005, and began the implementation of this programme among all staff and researchers using the facility. Audit inspections are conducted by AAALAC team upon accreditation, to assure compliance with all applicable occupational health and safety standards. The Department of Experimental Surgery, Singapore General Hospital was awarded full accreditation by AAALAC on 22nd June 2006, and maintains an occupational health and safety programme that is in line with AAALAC standards.

Setting Up an Occupational Health and Safety Programme

1. *Coverage*

   The DES Occupational Health and Safety Programme was set up to cover staff, scientists and researchers, students, visiting scholars, contractors and volunteers who have direct contact with animals housed within the DES animal facility, or indirect contact through exposure to unfixed animal tissues, fluids or wastes. The Programme also covers personnel who provide service support for animal-related equipment or carry out maintenance on building fixtures within the DES animal facility.

   All personnel enrolled into the programme have to undergo an orientation session and basic safety training. The type of safety training and information sheets issued are based on the level of exposure to animals, type of animal contact, and the individual's medical history. The DES Occupational Health and Safety (OH & S) team coordinates this risk assessment by working in partnership with professionals from the SGH Occupational Health and Epidemiology (OH & E) Unit.

   The OH & S team will send all identified individuals a standardized risk assessment questionnaire to fill out and return. The OH & S team will review all completed questionnaires and may, if deemed necessary, refer the individual to the SGH Occupational Health and Epidemiology Unit for further health assessment. Completion of the risk assessment questionnaire is required for the following personnel:

   - research personnel (both DES staff and external users) involved in animal research carried out at DES;
   - personnel involved in animal husbandry and the daily care of animals;
   - personnel working in laboratories where unfixed animal tissues from animals housed in DES are handled;
   - SGH staff who, as part of their normal job duties (e.g. administrative, housekeeping, maintenance staff) work in the building level that houses DES animal facilities;
   - external contractors and volunteers working within the DES animal facilities.

2. *Enrolment into the Occupational Health and Safety Programme*

   The process of enrolment varies according to the personnel involved. For new employees within the department, their respective supervisors are required to forward the names of the new employees to the OH & S team. In the case of new animal research projects, the Executive Research Coordinator at DES will gather the names of the all research personnel involved in the project at the time of IACUC application, and forward those names to the OH & S team. New research personnel joining an existing research project will be required to have their names submitted by the Principle Investigator, before starting work on the project.

These new personnel are required to fill in the risk assessment questionnaire, which are then reviewed and filed in the department. Review of risk assessment questionnaires will be conducted every three years. Individuals are required to contact the OH & S team if any changes occur in exposure levels or health status before the next scheduled review.

Visitors touring the facility and maintenance contractors are informed of the risks, instructed in the proper PPE for the area they will be in, and in most cases will be accompanied by a member of DES staff familiar with OHS SOPs.

3. *Responsibilities*
Both the Principal Investigator and the researchers/employees involved in the project have their responsibilities to ensure optimal occupational health and safety of all personnel involved.

The Principal Investigator/Supervisor
The responsibilities of the Principal Investigator (PI) or supervisor include ensuring that the eligible researchers/employees are enrolled in, and are in compliance with, the programme. PIs and supervisors are also responsible for the attendance of their personnel at mandatory orientation sessions and follow-up safety talks and training. They have to provide their personnel with time during the working day to attend orientation sessions or training dealing with the DES Occupational Health and Safety Programme.

PIs have to inform their personnel about occupational hazards (for example, the use of toxic chemicals in animal studies) being used in their respective animal research protocols. DES supervisors must notify employees of any possible exposure to hazardous biological, chemical, or physical agents in the workplace. All relevant personnel must be trained and have an acceptable level of proficiency when new equipment is to be used or new procedures are to be implemented. PIs and supervisors also have to provide adequate training and opportunity for hands-on practice if employees will be handling unfamiliar species.

Most importantly, all PIs and supervisors must report all animal bites or injuries received by their personnel to the DES OH & S team, and provide health insurance cover for their staff in the event of any injuries or accidents.

The Employee/Researcher
Employees and researchers have to inform their supervisor or PI of any animal bites, scratches, or injuries received or illnesses that may be related to working with animals. They also have to ensure that the supervisor/PI is informed of any work situation that might be hazardous. DES staff must notify the supervisor of equipment or facilities that are in need of repair.

All personnel working with animals have to use personal protective equipment and clothing, proper animal restraining devices and other applicable safety devices that are available. Good personal hygiene practices must be maintained while in the animal facility. In addition, researchers and employees must be familiar with all standard operating procedures for safety concerns and emergency situations.

All personnel in contact with animals must inform their primary care physician that their job responsibilities involve working with animals. The physician should be informed of the species handled, type of work involved and length of employment. This ensures that the physician is alerted to the possibility of zoonotic disease symptoms in the event of an infection.

Occupational Health and Safety Programme Components

Under standards outlined in *Occupational Health and Safety in the Care and Use of Research Animals*, the occupational health and safety Programme components must include: hazard identification and risk assessment; personnel training; personal hygiene; facilities, procedures and monitoring; personal protection; medical evaluation and preventive medicine. The components of DES Occupational Health and Safety Programme are outlined below:

1. *Hazard Identification and Risk Assessment*
   The Animal Facilities Biosafety Committee (AFBC) will evaluate animal research proposals on behalf of the SGH-IACUC. The hazard risks and recommended safety measures will be assessed based on the following guidelines:

   - The Radiation Protection Act 1992, regulated by the Health Sciences Authority (HSA), for the control and regulation of storage, use and disposal of radioactive materials.
   - The Environmental Pollution Control Act 1999, regulated by the National Environment Agency (NEA), for the classification of hazardous chemicals.
   - The Biological Agents and Toxins Act 2005, regulated by the Ministry of Health, Singapore (MOH), for the classification of hazardous biological agents and toxins.
   - The Workplace Safety and Health Act 2006, regulated by the Ministry of Manpower (MOM) Singapore, for workplace safety regulations.

     a) The Occupational Health Guidelines (Biomedical Sciences) issued by the Ministry of Manpower (MOM) Singapore.
     b) The Singapore Biosafety Guidelines for Research on Genetically Modified Organisms 2006, issued by the Genetic Modification Advisory Committee of Singapore (GMAC).
     c) SGH Ethics Committee/Institutional Review Board (IRB) and the SGH Institutional Biosafety Committee (IBC) for the safety and legal issues involving the use of human derived tissues and extracts.
     d) Guidelines from the Agri-food and Veterinary Authority (AVA) for animal disease surveillance and importation of laboratory animals from accredited suppliers.

Once the protocols have been approved, the AFBC will inform SGH-IACUC. Protocols are eligible for SGH-IACUC approval only after obtaining approval from the AFBC.

In addition to National Acts and Guidelines applied to animal research protocols, DES also monitors risk for staff and researchers using the animal facilities. All personnel enrolled in the Programme have to fill out the DES risk assessment questionnaire, as detailed under the section on "Coverage". If any hazardous risks (biological, chemical or physical agents) or potential health risks (allergies and asthma) are identified, the OH & S team will refer the individual to the SGH Occupational Health and Epidemiology Unit for a follow-up health assessment.

The SGH Staff Clinic also conducts pre-employment health screenings of all new employees to assess potential health risks. The pre-employment medical check-up includes immunization against Hepatitis B and tetanus, and screening for tuberculosis.

2. *Personnel Training*

A department veterinarian will prepare information sheets detailing zoonotic diseases in each animal species and their manifestation in infected animals. These sheets will be issued to personnel exposed at a high level to the animals involved. The veterinarian will also give safety talks on zoonotic diseases that lead to illness in humans. The veterinarian, in collaboration with the OH & S team, will determine whether any additional talks or safety training is required, and will provide the necessary talks or training to personnel involved.

It is mandatory for all personnel working with radioactive materials to attend the Basic Radiation Safety Awareness Course conducted by the Department of Nuclear Medicine. Personnel who work with chemicals in the lab are also trained to identify hazardous, corrosive and flammable chemicals and to store them only in designated cabinets. These staff must refer to the MSDS when in doubt, and are required to know how to use the chemical spill kits in the event of chemical spills. All personnel who are required to wear the N95 facemask (as assessed with the risk assessment questionnaire) are trained on proper use of the N95 facemask through a mask-fitting course conducted by the OH & E Unit.

All DES staffs are trained on the use of fire fighting equipment conducted by the Department of Facility & Plant Engineering. Annual drills are conducted to train staff on the SGH Emergency Response Plan in case of fire outbreaks.

3. *Personal Hygiene*

All personnel are not permitted to eat, drink, put on contact lenses or apply cosmetics in the animal facility. Hands and fingers should be kept away from the mouth, eyes, nose and hair after handling animals. Hands must also be washed with disinfectant soap and water after handling animals or their secretions and excretions, even if gloves were worn.

Personnel should also avoid working with animals when ill, especially with respiratory symptoms. This is to prevent infectious agents spreading between the personnel and the animal, and vice versa. Additional precautions must be taken with open wounds by covering up the wound with a water-resistant band-aid.

4. *Facilities, Procedures and Monitoring*
Facilities used for animal experimentation with hazardous agents are designed to be separate from other animal housing and support areas. Personnel are to note that hazardous agents should be contained within the study environment (for example, a Biosafety Cabinet), which must be designed with appropriate barriers against accidental release or escape of these agents. Biosafety Cabinets are necessary and must be used for protocols that involve the handling of infectious or toxic agents. In the event of animal surgeries and anaesthesia, waste anaesthetic gases are extracted via the wall mounted scavenging port and are discharged to the exterior using a vacuum pump located in the Level 3 pump room.

Proper signage must be displayed for specific hazards on the entry door and in many cases the individual cage. Signage should have the appropriate hazard symbol, the name of the hazardous agent/material, a summary of the risk, contact information for responsible persons, safety precautions, appropriate PPE to be used, information on handling of animals, cages and waste, and what to do in the event of exposure.

Differential air pressure is used to control the flow of air in the rodent caging system. Positive differential air pressure of vented cage racks in the Specific Pathogen Free (SPF) "clean" room is used to ensure that contaminated environmental air is prevented from flowing into the cages, and keeps the animals free of the specified pathogens. Negative differential air pressure of racks in all the other rodent rooms ensures that the air from the cages do not leak into the room environment and lowers the risk of zoonotic infections among researchers exposed to contaminated cage air. The differential air pressure in the ventilators are monitored daily.

Negative differential air pressure is also maintained in the Virology Laboratory to ensure that the contaminated air from inside the lab is prevented from flowing out to the common corridor. The differential air pressure is monitored by an in-built pressure monitor and maintained by a door lock system.

5. *Personal Protection*

Personnel are required to remove their laboratory coats and overalls before entering the animal holding areas. They must be fully gowned up in standard Personal Protection Equipment (PPE), which include disposable gowns, masks, head covers, shoe covers and gloves. Personnel with potential exposure to hazardous agents are to be provided with additional PPE, as outlined below.

Personnel exposed to nonhuman primates must wear the N95 mask instead of a surgical mask and a splash shield to avoid any exposure of mucosal surfaces. Personnel handling X-ray materials and equipment or the fluoroscope are to wear lead aprons and thyroid guards. Those exposed to beta radiation are to be trained to exercise appropriate precautions such as working behind acrylic shields. All personnel handling any form of radioactive materials must wear their personal dosimeters issued by the Health Sciences Authority Centre for Radiation Protection.

Personnel working in designated biosafety areas where they might be exposed to airborne agents are to wear the N95 mask instead of normal surgical masks. The Occupational Health and Epidemiology Unit will issue personnel with animal allergies special respirators upon recommendation after the risk assessment exercise. The same applies to personnel with asthma who are required to wear the N95 mask.

Lastly, all personnel are required to wash their hands and dispose used PPE at a designated waste trolley after the animal procedure. Used PPE are not to be worn outside the animal facilities.

6. *Medical Evaluation and Preventive Medicine*

All animal care personnel will be immunized against Hepatitis B once every 5 years, and against tetanus once every 10 years. The Mantoux test will also be administered once every year to monitor for tuberculosis.

Pre-employment screening is followed one month later with a post-employment check-up, where the OH & E physician can assess the animal exposure level of the individual. After the post-employment check-up, all personnel will undergo an annual medical screening as a form of medical surveillance. Personnel have to fill in the Occupational Health Review Authorisation issued by the OH & E Unit prior to their regular screenings.

Personnel must notify their respective supervisors of possible exposures from animal bites, scratches, needle-prick injuries, other work-related illnesses or accidents, and allergies. All personnel, especially those handling nonhuman primates, are to be given clear instructions to easily locate and use bite, scratch and splash-care stations. The department OH & S team will be notified by the supervisors in cases of trauma, mucosal or sharps exposures and liaise with the OH & E Unit. Personnel must then proceed to the relevant clinic immediately for treatment.

A standard protocol for the management of bite, scratch or splash incidents involving non-human primates are clearly defined for the attending OH & E physician at the clinic. The attending physician will implement the treatment and conduct follow-up monitoring for the personnel involved.

CHAPTER **6.2**

NEW EMPLOYEE AND EXTERNAL USERS ORIENTATION

Inria Kurniawan Then

The orientation of new employees and external users is an important component of the Occupational Health and Safety (OH & S) programme of an animal research set up. This has implications on biosafety issues pertaining to zoonoses, allergens, chemical and radioactive hazards. It is therefore critical that new users of animal facilities, be it new employee or external user, be subjected to strict orientation for early enforcement of personal safety as a preventive protection against unnecessary incidents. Each animal research centre will has its own implementation system and this chapter describes what is being practiced in Department of Experimental Surgery (DES).

DES actively collaborates with both local and overseas research and tertiary institutions in collaborative and work attachment programmes. Research collaborators will include clinicians, scientists, and research associates. Work attachment programmes cover students and institutional trainee staff. They are classified as external users of research and animal facilities in the department. All external users have to undergo an orientation programme to familiarize them to the work environment and be in compliance with department specific safety rules and regulations before they are allowed to conduct any work at the department.

New research collaborators can be either team members of new projects or additional personnel for on-going projects that have been approved by SingHealth IACUC. Attachment programmes will cover individuals or groups that come for training attachment, industrial attachment, work attachment, internship or purely as observers. These are arranged between institutions or via self-referrals. Depending on the type of attachment and external personnel involved, preliminary agreements and documents have to be approved to meet the requirements of Singapore General Hospital policy on work attachment. These processes are being coordinated by various parties in SGH, such as Human Resource, Postgraduate Medical Institute, Associate Dean's Office or by the department itself.

External Users Orientation

Before a compulsory general orientation programme is conducted, the following forms or documentation must be submitted to DES:

1. *Principal Investigator (PI) Declaration Form*
   The PI must declare the contact details of team members who will be accessing DES premises, and list of all equipments, reagents and chemicals to be brought into the premises. MSDS has to be provided for hazardous substances. PI must also declare that he/she will be accountable for all team members' compliance with IACUC and departmental regulations during the course of study in DES and ensure safety of all personnel involved in the project. Any change of personnel and inventory update must be communicated to DES.

2. *Occupational Health and Safety (OH & S) Risk Assessment Form*
   Each team member must submit a completed OH & S Risk Assessment Form. The form will be evaluated once the research work commences, and review of risk assessment questionnaires will be conducted every three years. Individuals are required to contact DES OH & S team if any changes occur in risk exposure levels or health status before the next scheduled review.

3. *Undertaking Letter*
   This letter certifies that the individual has read and understood the health and safety issues as listed in SGH standard operating procedures.

4. *Letter of Indemnity*
   SGH and DES require indemnity from the external personnel for any accident, mishap or injury suffered by them during the course of work in DES. DES also needs verification that all personnel accessing its premises hold insurance coverage for work-related incident.

5. *Non-Disclosure Agreement (NDA)*
   Personnel on attachment have to sign additional Non-Disclosure Agreement (NDA). The NDA ensures that they will not disclose or duplicate confidential information for any purpose other than for approved collaborative activities. Prior written consent has to be obtained from SGH before any disclosure to outside parties.

The DES General Orientation Session for external users is conducted by the department supervisor or Institutional Veterinarian. It aims to familiarize all external personnel to relevant procedures involved in biomedical research and ensure their awareness of workplace health and safety issues. The orientation programme includes briefing session that highlights important issues in doing work at DES to be followed by a familiarization site tour. Before the orientation, DES examines the project requirements, level of animal exposure of each personnel and certification for conducting specific procedures or specialized equipment operation.

The orientation session covers the following:

1. *Animal order and quarantine procedures*
   All external users bringing animals into DES facilities have to seek approval by filling up the prescribed form with one-week advance notice. Animals on arrival have to be accompanied by health certificate and will have to be acclimatized prior to use. DES will not accept any return of animals that have been brought out of DES facilities. The PI must ensure that the number of animals used does not exceed the number approved by IACUC.

2. *Facility access procedures*
   External users are informed on the official operating hours and approval process for activities outside the working hours. The external personnel have to notify DES in advance for use of the facilities or submit the project schedule to the department. DES reserves the right to refuse any entry of unauthorized personnel, whose name not listed in the PI Declaration Form, or external users who wish to enter into the facility without prior arrangement.

3. *Traffic flow guidelines and requirement of Personal Protective Equipment (PPE)*
   To prevent cross contamination, all external users have to follow designated traffic flow plan and adhere to recommended protective clothing for each specific animal holding areas. All external users are made aware of the type of facilities and its environmental requirement and access restriction. Appropriate use of PPE is compulsory in the animal facilities.

4. *Cage use and transport*
   External users have to be responsible for compliance to animal holding regulations and are required to use of appropriate transport containment and its return to cage washroom for sanitation after use.

5. *Animal necropsy*
   External users are briefed on appropriate methods for animal necropsy especially those pertaining to the use of microbial kill tank and disinfection of the working area and instruments will also be highlighted.

6. *Disposal of animals*
   External users are informed on the requirement of double bagging of carcass and unwanted tissue, and the location of disposal sites. The use of appropriate colour-coded bag for waste disposal (yellow: biohazardous, red: radioactive, purple: cytotoxic, black: general waste materials, orange: soiled linen) is also emphasized.

7. *Sanitation and personal hygiene*
   All individuals are responsible for maintaining of clean work environment and should practice personal hygiene at all times.

8. *Handling of hazardous substances*
   Importance of proper containment, storage, labeling and transportation of potentially hazardous substances are highlighted in this section. Safety recommendations from MSDS, SingHealth IACUC or other relevant safety regulating committees are to be executed properly.

9. *Overview of DES's emergency preparedness, planning and response programme*
   The programme covers security surveillance, disaster prevention and preparedness, disease outbreak, animal escape, and workplace accident. The external users are briefly introduced to the programme and issued with emergency contact number.

10. *Overview of DES's occupational safety programme*
    The external users are informed on the hazard identification and safety assessment processes, PI responsibilities, project-related personnel training requirement, medical surveillance, and incident prevention and reporting.

11. *Animal facilities floor plan*
    The external users are given overview of DES floor plan and locations of first aid kit, eyewash and shower stations, spill kit, fire hose reel, evacuation routes, disposal site and other relevant sites.

12. *Zoonotic diseases*
    DES veterinarian provides safety talks on zoonotic diseases relevant to the research project. Information sheets detailing zoonotic diseases in each animal species and their manifestation in infected animals will be issued to personnel exposed at a high level to the animals involved.

13. *Site tour*
    Site tour is conducted after the briefing to familiarize external users with facilities segregation for different biosafety level, room and equipment function and usage restriction.

14. *Specific training*
    DES will conduct additional project-specific training as required, especially for projects, which involve DES technical collaborators.

New Employee Orientation

DES orientation programme for new employees is more comprehensive than that for external users. The new employees will undergo several stages of orientation before their actual commencement of work activities. They are administrative orientation programme, departmental general orientation programme, hospital orientation programme, job specific and continual training.

Firstly, to facilitate the settling-in of the new employee at the department, he/she will be assigned a "buddy", who is a senior staff member, to ensure smooth transition into the new work environment. Essential resources will be provided, which include workstation, computer, department keys and access card. Basic information relevant to the job, key responsibility, continual training programme, performance expectation and duty rosters are also explained.

The administrative orientation programme training will either be through oral communication conducted by senior staff members or as reading assignments on text documents. The topics covered include:

1. DES's administrative coordination and management programme, which describe the complete overview of work processes in DES. The new employee will also be introduced to laws and guidelines in biomedical research and overseeing ethics committees (for example, IACUC, IRB and AFBC).
2. DES's OH & S programme with emphasis on hazard identification and safety assessment processes, supervisor responsibilities, project-related personnel training requirement, medical surveillance, and incident prevention and reporting.
3. DES's programme for emergency planning and response, to ensure the new employee is well informed on the availability of proper measures in handling emergency situation and expected response should any incident happens.
4. Overview of DES's standard operating procedures, veterinary care programme, *in vitro* laboratory and chemical management programme, and laboratory safety manuals. Where applicable, the new employee will be instructed to familiarize himself/herself with specific work processes relevant to her/his duties.
5. Brief overview of hospital related procedures and policies, such as Human Resource Policy, Quality Management System (ISO9001), Environment Management System (ISO14001), Occupational Health and Safety Assessment Series (OHSAS18001).

General orientation programme follows that for new external users. DES will assign senior staff as the mentor to the new staff to provide job-specific training. The new employee will also be automatically enrolled in Responsible Care and Use of Laboratory Animal Course, Radiation Safety and Awareness Course, Infection Control and Mask Fitting Training Course. He/she will be scheduled to attend Hospital-level orientation in the next immediate session to have clear understanding of hospital core values and missions.

CHAPTER 6.3

RADIATION SAFETY AWARENESS IN ANIMAL RESEARCH

S. Somanesan

Radioactivity

Certain nuclides are found to be unstable as they occur in nature. These are called natural radionuclides. Examples of these are ^{238}U and ^{226}Ra. Natural radiation is all around us and cannot be avoided. We receive our personal radiation dose from cosmic rays arising from solar flares in outer space, Gamma rays from the earth, floors, building materials and radon gas emanating from rocks and soil. We also ingest natural radioactivity through our diet. Certain food like nuts and coffee concentrate a higher radioactivity.

A. *Types of Ionising Radiation*
   Table 6.3.1 shows the different types of ionising radiation.

Table 6.3.1: Types of ionizing radiation

Type	**Alpha Particles (α)**	**Beta particles (β)**	**Gamma radiation (γ) & X-ray**
Form	Particulate form These are Helium nuclei	Particulate form Either electrons or positrons	Photon of electromagnetic radiation (EMR)
Ionising ability	They are highly ionising	Less ionising than alpha	Low ionising ability
Penetration range	Low penetrating range Stopped by few cm of air	Low penetrating range Stopped by plastic	Highly penetrating range Stopped by lead or heavy metals
Stored in	Plastic containers	Plastic containers	Lead containers
Detected by	Semi-conductor devices	Geiger counters can be used	Scintillation and Geiger counters

B. *Activity*

Activity can be defined as a measure of radioactivity that is proportional to the number of nuclear transactions/time, that is $A = dN/dt$ where N and t refer to the number of particles present and the time taken. The SI unit of Activity is the Becqueral (Bq). 1 mCi = 37 MBq.

C. *Biological Effects of Radiation*

For a given dose, the biological effects depend on the type of radiation and the type of tissue being irradiated. Different organs of the body vary in their sensitivity to an absorbed dose of radiation. Organs with a high rate of cell replication such as bone marrow, lung, thyroid, gonads and the female breast are the most sensitive to radiation damage. The various types of ionising radiation (α, β particles, Gamma and X rays) have differing ionising ability and hence differing weighting factors.

D. *Absorbed Dose, D and Dose Dose Equivalence, H*

Measure of the energy (E) imparted per unit mass (m) of tissue, that is, Dose, $D = E/m$. 1 Gray (Gy) = 1 J/kg. Dose equivalence was defined to compare and equalize the effects of different radiation. Hence $H = D \times W$ where W is the weighting factor. Unit is the Sievert, Sv. 1 mSv = 100 mrem = 0.1 rem and 1 mrem = 0.01 mSv. This is a more important term in radiation protection. Our annual radiation dose limits are based on this term.

E. *Sources of Background Radiation*

The background radiation arises from natural radioactivity in the air, ground, food and drinks that are consumed, cosmic radiation, air travel, nuclear power plants, nuclear weapon testing and medical sources. Cosmic radiation increases with increasing altitude. The exposure rate is about 5 μSv/hr at 10 km above the earth, where most commercial flights fly and is about 0.03 μSv at sea level.

Radiation Safety Programme

The rationale of the radiation safety programme is to prevent individuals, the environment and future generations from being subjected to unacceptably high radiation exposure. The programme rests on the International Commission of Radiation Protection (ICRP) principles of:

- **Justification:** where the benefit should strongly outweigh the use of radiation;
- **Optimization:** where the dose must be kept as low as reasonably achievable; and
- **Limitation:** which refers to the exposure to be kept below annual limit.

A. *Occupational Radiation Exposure*

ICRP recommends a whole body radiation exposure of 20 mSv/year for radiation workers. The annual limit for extremity doses such, as that for fingers is 500 mSv. Several countries, including Singapore, incorporate these recommended dose limits in their Radiation Protection Acts and Regulations. The occupational dose of radiation workers in Singapore is monitored by the Centre for Radiation Protection, (CRP) of the Health Sciences Authority (HSA). The records of monthly, annual as well as the lifetime doses received by radiation workers are kept at the CRP. These values are determined by thermoluminescent dosimeters (TLDs) worn at the waist level. They are lithium fluoride (LiF) crystals within a plastic holder and are able of registering the received amount of radiation exposure. TLDs are able to record total exposures and they are returned to CRP for dose assessment. Pocket dosimeters are electronic devices that provide real time radiation exposures readings. Radiation levels in the laboratory and imaging rooms where radioactivity and radiation are used will be higher than the background levels.

B. *Foundation for a Safety Programme*

Radiation exposure to personnel can be reduced by decreasing the time of exposure to a radiation source and by increasing the distance and shielding between the radiation source and the personnel.

Time

- Reducing the time spent with ionising radiation can reduce radiation exposure.
- Laboratory or animal work with radionuclides has to be done expediently and efficiently to reduce the radiation exposure.

Distance

- Increasing the distance from the radiation source can reduce radiation exposure.
- Radiation obeys the inverse square law. If the distance between the radioactive source in a vial and the laboratory staff is doubled the dose rate is reduced by a factor of four.

Shielding

- Shielding a radiation source with an appropriate material such as lead, concrete or plastic in the case of beta (such as P32) and alpha sources can reduce radiation exposure.

C. *Internal Contamination by a Radionuclide*

Internal contamination by a radionuclide is possible by three routes. They are namely,

- penetration through skin (such as injuries by contaminated blades and needles);
- accidental ingestion; and
- inhalation via aerosol.

The routes of entry can be avoided with responsible and good laboratory practices.

Use of Radionuclides in Animals

When animals are used as experimental tools in studying biological processes, attention is required on the waste disposal and handling. Radioactive waste from living tissues appears as a solid, liquid, and gas. All experimental animals treated with radionuclides must be considered as radioactive waste at the termination of the experimental procedure. Extreme care must therefore be taken in the experimental design to ensure that contamination does not occur. All animals with radionuclides should be isolated in cages or leak-proof containers labeled with the radioactive warning hazard symbol. The bottom of the cages should be lined with absorbent paper pads with plastic backing to prevent contamination. Excreta must be collected and properly stored and monitored prior to its disposal in the sewer. Dead animal must be preserved so as to prevent bacterial or fungal decomposition while awaiting the physical decay of the radioactivity. Attending staff should use disposable gloves, gowns and shoe covers (where appropriate) when handling radioactive waste and the animals. Cages are to be cleaned and surveyed at the end of each individual experiment and exposure readings documented.

Workplace Radiation Monitoring

The practice of monitoring the workplace ensures that the radiation exposure to laboratory staff and the public are kept low by minimizing exposure and contamination. This practice ensures a safe work area. The frequency of workplace radiation monitoring is based on expected changes in the radiation environment (for example, daily, weekly). In a laboratory where radionuclides are used, workbenches, storage areas, floors, animal holding areas, injection and imaging rooms need to be surveyed for radiation. Surveys should also be conducted in radiation isolation rooms where animals are undergoing radionuclide therapy as well.

Summary of Good Laboratory Practices

Below is a summary of good laboratory practices when working with radionuclides or in an environment where radionuclides experiments are conducted.

- No food may be eaten or stored in the laboratory.
- No liquids should be consumed or stored in the laboratory and water from the sink is not to be consumed.
- Cosmetics should not be applied and smoking should be prohibited in the laboratory.
- Pipetting by mouth of any liquid containing a radioactive substance is strictly forbidden.
- Wear proper personal protective gear (PPE) which includes a laboratory coat, double disposable latex gloves. Shoe covers, head netting and facial mask are worn only if necessary.
- Radionuclide users will have to wear ring and whole body dosimeters.

- Besides proper PPE such as disposable double gloves and gowns booties need to be worn in the injection/infusion room and in the animal housing room if there is a significant potential for the floor to become contaminated. Booties will be removed and disposed of as radioactive waste before the individual leaves the potentially contaminated area.
- After handling contaminated animals, bedding, or cages, researchers and animal handlers will monitor their hands, arms, clothing, and shoes for contamination. Any detectable contamination must be cleaned immediately.
- Radionuclides have to be transported in appropriate shielded containers.
- A waterproof plaster must cover all wounds and abrasions before entering the laboratory.
- Female radiation workers must inform their supervisors when they are pregnant.
- All spillage or suspected spills must be reported to the safety officer and/or the Principal Investigator (PI).
- Radioactive waste must only be placed in appropriately shielded containers.
- Absorbent paper will be placed underneath the animals throughout the injection or infusion procedure besides under the cages in the housing area to prevent the floor or any other surface from becoming contaminated.
- Only one animal can undergo a radiological or radionuclide imaging in a room at any one time. This is to reduce the radiation exposure to the users.
- An animal shall not be held by any individual during a radiological or radionuclide scan unless other means of immobilization are impracticable. If manual restraint is necessary, the animal shall be held down by a minimum number of individuals who are not staff of the veterinary establishment, not pregnant and not below the age of 18 years. If it is necessary for the animal to be held by members of the veterinary establishment, only those who have been registered as radiation workers and have been trained for such purposes shall be so employed and they shall be provided with protective clothing and be positioned so as to avoid the primary beam.
- Before a necropsy is performed on an animal, which has received an injection, or infusion of radioactive material, the Radiation Safety Officer will determine what, if any, radiation safety precautions are necessary.
- Users of high-energy beta or gamma nuclides should wear eye protection, such as safety glasses or work behind a barrier if there is a possibility of a spill or if working in close proximity (< 15cm to the face).
- The imaging table and laboratory benches have to be completely covered with disposable absorbent pads with the absorbent side up. The animal will be placed on top of these pads for the duration of the study.
- Radioactively soiled cages have to be stored behind appropriate shields for decay.
- Biological waste (including blood, tissues, and carcasses) has to be placed in plastic bags and the bags sealed and appropriately labeled before being placed in a "Radioactive" freezer for decay prior to disposal.

Radioactive Spills

In case of radioactive spills, proper methods and procedures must be followed to contain the spills. Below are the steps involved in handling radioactive spills:

- Wear appropriate PPE.
- Perform a radiation survey and mark out the contaminated area.
- Cover area with absorbent paper towels to soak up the liquid.
- Absorb the liquid spill with absorbent paper towels without unduly increasing the spill.
- Wipe the spill in an inward direction to prevent increasing contaminated area.
- Clean the spill site with a good soap solution in an inward direction to prevent increasing contaminated area.
- Monitor radiation levels of all contaminated gloves, paper towels, cleaning instruments and dispose them as solid radioactive waste.
- Spill area is to be monitored for the presence of radioactivity. Marking is to be removed upon complete removal of contamination.

Radioactive Waste

Radioactive waste must be handled safely and properly bagged to provide a safe and hygienic environment as well as to prevent radioactive contamination and unnecessary radiation exposure to laboratory and housekeeping staff. Radioactive waste refers to solid and liquid waste that is contaminated with radionuclides. Solid waste is further segregated into sharps and non-sharp waste.

1. *Radioactive Waste Disposal of Short Half-Life Radionuclides*
   Solid radioactive waste has to be placed in red radioactive disposal bags with the radiation hazard symbol and stored in a shielded area for a minimum of 10 half-lives. When the radiation levels reached 0.2 mR/h a disposal form is filled and sent to CRP, HSA. Upon CRP's endorsement, housekeeping department engages a licensed private contractor to remove and dispose the waste at a NEA approved land or marine fill. Sharp waste that includes spent syringes and needles are biohazaradous as well. These wastes have to be stored in yellow plastic biohazaradous containers behind an appropriate radiation shield. Once filled they are closed, dated and stored for a minimum of ten half-lives. Disposal is only possible with CRP's endorsement.

2. *Liquid Radioactive Waste*
   Liquid waste has to be stored in an appropriate container until its radioactive level is acceptably low enough for disposal. The maximum allowable concentration of radioactivity that can be discharged into the sewer for each radionuclide is available in the Radiation Protection Regulation of Singapore. The radioactive concentration of the liquid waste can be determined with a gamma or a beta counter. It is then discharged into the sewer with diluting. Records are kept of each discharge. Some

maximum allowable radioactive concentration for common radionuclides to be discharged in water is as follows:

^{3}H	$1 \times 10^{-2} \mu Ci/ml$	^{125}I	$4 \times 10^{-6} \mu Ci/ml$
^{14}C	$2 \times 10^{-3} \mu Ci/ml$	^{51}Cr	$5 \times 10^{-3} \mu Ci/ml$
^{32}P	$5 \times 10^{-5} \mu Ci/ml$	^{35}S	$2 \times 10^{-4} \mu Ci/ml$

3. *Labeling Requirements*
Work areas including cages used to transport and/or house the animals with radionuclides in them have to be labeled "Caution Radioactive Materials" hazard signs. Each room in which radioactive materials are used must bear a label "Caution Radiation" on doors to the room. These labels must have the radioactive hazard symbol.

CHAPTER 6.4

EMERGENCY CRISIS MANAGEMENT

Irene Kee

Implementation of Emergency Crisis Management is dependent on the Animal Research Institution set up. Some are standalone entities that must rely totally on their own resources. Others are part of a campus system such as a hospital or university, allowing them to tap into shared resources.

The description in this chapter is based on experiences of Department of Experimental Surgery, Singapore General Hospital (SGH) and will be useful as a guide for implementation of emergency crisis management in other animal research institutions.

The Department of Experimental Surgery (DES) mandates an emergency crisis plan that is in accordance with the Hospital (SGH) policies on Emergency Preparedness and Response Plan and the Business Continuity Plan as defined by its Safety Network Committee. Additional in-house policies and processes pertaining to animal research facility are developed in order to be in compliance with the legal guidelines of the National Advisory Committee for Laboratory Animal Research (NACLAR).

On 1 April 2005, a committee for Emergency Crisis Management of Animal Facilities was appointed to oversee the formulation of an action plan for workplace accidents, infrastructure breakdown, external factors, compliance to safety regulation, prevention and containment measures for zoonotic disease outbreak and prevention of animal escape.

There are 5 areas of structural processes pertaining to the Emergency Crisis Plan for the Department of Experimental Surgery, namely, i) communication outreach and control, ii) disaster prevention and preparedness, iii) disease outbreak containment, iv) animal escape and v) workplace accidents.

Communication Outreach and Control

The objective of Communication Outreach and Control is to address public concerns on animal research and minimizing the risk of potential break-ins, harassment by mail and other mischief by groups opposed to animal research. Public support is dependent on ensuring compliance and maintaining an animal care and use programme of impeccable integrity. All research activities are to be strictly confidential and the Institutional Animal Care and Use Committee (IACUC) will review and be responsible for assessing the vulnerability of all the protocols, for example, neuroscience, drug addiction or toxicology studies. Passwords have to be chosen wisely to safeguard sensitive data and personal information with all data being back up on tape, diskette or other media. Security checks on PCs and notebooks/laptops are conducted by SingHealth IT and any user found engaging in prohibited actions that compromised the information security will have their access and usage of SingHealth's information resources temporarily or permanently suspended.

Disaster Prevention and Preparedness

The aim of Disaster Prevention and Preparedness is to minimize the negative impact of utility and communication disruptions in the event of an unexpected crisis situation (e.g., natural disasters). In the event of any disruption to the power supply, there are two backup generators to power the entire hospital with sufficient diesel fuel supply to power the generators for three days. SGH has water reservoir storage supplied directly from PUB for the hospital's usage for three days if there is any disruption to the water supply. For disruption to water supply of more than three days, the reservoir will then by topped up by water supplied through water tankers. The hospital currently provides the PABX system with 4000 internal PABX lines throughout the hospital. About 1000 handphones have been issued to hospital staff to allow for efficient communication in the events of any disruption to Telecommunication (PABX) services.

Currently, SGH is equipped with five fire suppression systems namely smoke detector, fire alarm system, hose reels, fire extinguishers and FM200 for data centre. Staffs are trained through fire drill exercise to follow the sequential steps under the following headings:

R (Rescue)	Rescue or move personnel away from fire
A (Alarm)	Alert staff and raise alarm by breaking glass of alarm call point
C (Call)	Call Fault Reporting Centre and provide location and extent of fire
E (Extinguish)	Try to extinguish the fire with portable fire extinguisher (Fig 6.4.1) or hose reel without personal risk
S (Shut)	If fire is beyond control, shut down gas supply and close the door to contain the fire and smoke and follow the evacuation procedures

Fig 6.4.1: Fire extinguisher.

Staffs are to evacuate the building in an orderly manner following the specified evacuation routes, along the road leading to the open field behind the designated assembly point, the Department of Pathology building.

For DES satellite facility, the Animal Husbandry & Hospital in Sembawang, a backup generator was installed to provide emergency power in the event of a power failure. Water tanks are provided to supply drinking water to the animals via pressure pumps, which are also connected to the standby generator. If the pump that direct wastewater discharge from the waste collecting well to the waste lagoon fails, there is provision for an overflow mechanism in the well that will direct water through gradient flow.

Disease Outbreak Containment

The objective of disease outbreak containment is to provide an emergency response plan in the case of a disease outbreak occurring in the animal holding area and report formally to IACUC, AVA and SGH management.

Staffs and researchers should report any abnormal behaviour, ill health or any unexplained sudden or mass death of animals to the Veterinarian as soon as possible. In the event where an animal is suspected to have died from a contagious disease, biosafety procedures must be observed during necropsy which includes donning of full Personal Protection Equipment (PPE) and N95 mask. All tissue samples must be contained in sealed tubes, placed in double Ziploc bags and sprayed with disinfectant before they are allowed to be taken out for testing. All rodents in the room, where a contagious disease outbreak is confirmed, may be culled by carbon dioxide inhalation. The animal carcasses will be deposited in Ziploc bags and immersed in 10% formalin. Large animals will be euthanized by anaesthetic overdose and deposited directly into the biohazard bin, which will be secured and then send for incineration on the same day.

Virkon S (1:100), a sterilizing agent containing a balanced blend of peroxide will be used to disinfect all contaminated equipment, cages, ceiling and floor. Contaminated rooms will be cleansed thoroughly with 1 % bleach (sodium hypochlorite) and then sealed off for a period of two weeks. Any contagious disease outbreak that involved an infectious agent that threatens human health is to be reported immediately to the IACUC, AVA and SGH management. The Occupational Health & Epidemiology Unit will provide the appropriate prophylactic treatment to personnel who might have contact with the infected animals.

Animal Escape

The animal facility must be a secured area with access restricted to authorized personnel only. The animal holding rooms have escape-proof ceilings constructed of bonded calcium silicate and air vent diffusers are secured with screws. The animal cages are incorporated with escape-proof features with appropriate sizes of mesh and locking mechanisms to keep the animals in. All animals in the animal facility are accounted for on a daily basis and a weekly inventory report. Transport boxes or trolleys with lockable latches are used for transit of animals and containment of animals is further enhanced with a double locking system. Transport vehicle will have a secondary barrier with either paneled metal sheet or a meshed wire enclosure to contain animal escape during transit.

Workplace Accidents

The objective of a standard operating procedure for workplace accidents is to generate awareness among staff to workplace hazards and understand processes to be observed in the event of accidents. This is to minimize risk of accident occurrence with a support plan for immediate response to contain dangerous effects. Measures adopted should be in compliance to the legal framework of the Workplace Safety and Health Act.

Personnel are to don protective attire, including aprons and eye protection when handling hazardous chemicals or solvents, which are corrosive, toxic flammable and explosive. All manipulation involving hazardous chemicals or solvents are to be carried out in the fume cupboard. All flammable and explosive chemicals are to be kept away from all heat and ignition sources especially naked flames. Extreme caution should be exercised when handling spillage of corrosive, flammable or explosive substances. In the event of spillage of corrosive or explosive substances, acids, alkalis or any explosive substances should be neutralized before clean up is attempted with paper towel, a non-flammable detergent, sand and other absorbent material (Fig 6.4.2).

Another activity that might contribute to workplace accidents is the use of radioactive substances. All radiation workers have to be registered with the Health Sciences Authority Centre for Radiation Protection. These registered staffs must wear the radiation monitoring devices such as personnel thermoluminescent dosimeter at all times when dealing with radiation work. All radionuclide work must be carried out behind a lead shield for gamma rays and a lucite shield for beta emission. The Radiopharmacist, Radiation Physicist or Radiologist will perform animal procedures, which involve injection of radioactive substances. However, they may assign competent technicians to carry out these tasks. In the case of radioactive materials spillage, the affected area is to be marked out clearly with tape and the radiation hazard symbol to prevent contamination by unsuspecting persons. The international radiation hazard symbol must be displayed prominently outside all rooms where ionizing radiation work is carried out.

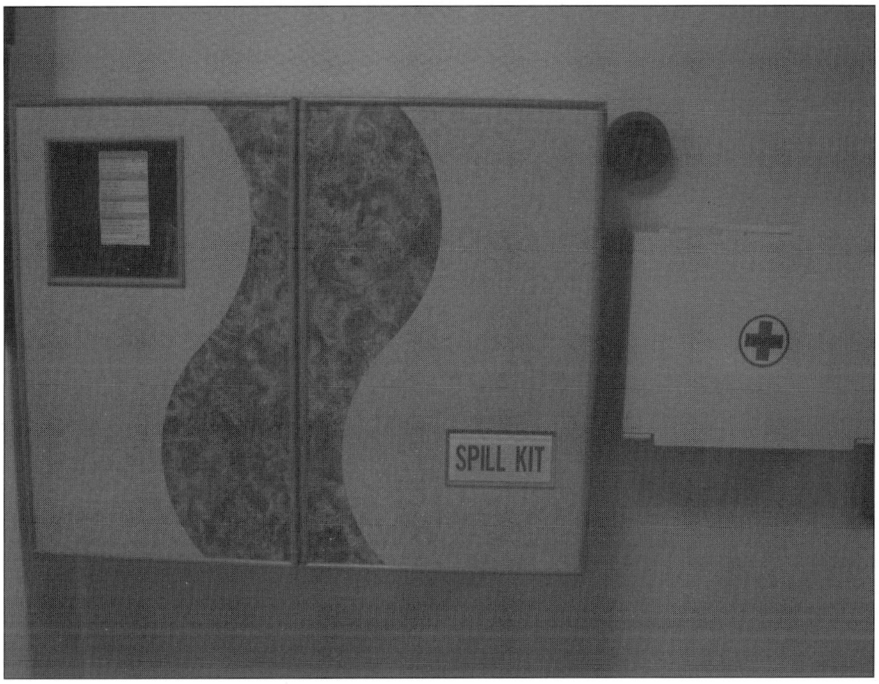

Fig 6.4.2: Spill kit and first aid kit box.

CHAPTER 6.5

ZOONOSES AND LABORATORY ANIMAL ALLERGIES

Jason Villano

The occupational health and safety programme in a laboratory animal facility should ensure that the risks associated with the use of these animals are reduced to acceptable levels. Potential hazards — like zoonotic agents and allergens — inherent in or intrinsic to animal use should be identified and the risks assessed.

Laboratory Animal Allergy (LAA)

Laboratory animal allergy or LAA occurs as an altered reactivity following second or subsequent exposure to an allergen in a laboratory animal facility. It develops in approximately 10 to 30 % of the personnel who work with laboratory animals, making it one of the most common occupational health problems. Meanwhile, people who are prone to allergies have an approximately 70 % chance of developing LAA if working unprotected with rodents.

The symptoms of LAA are usually controllable but may become severe if preventive measures are not in place. Inhalation of allergens can result in sneezing, itchy eyes and asthma while direct skin contact can result in itching and localized swelling of the skin. In extreme cases, anaphylaxis can develop and produce life-threatening consequences from laryngeal oedema, airway obstruction and shock in certain individuals with severe reactivity. The risk factors of LAA include the potency of the allergen, genetic predisposition, pre-existing allergy to other agents and factors that relate to exposure. Although controversy surrounds the effects of tobacco smoking in the development of LAA, studies have shown that it increases serum levels of immunoglobulin E (IgE) antibodies such as those produced in immediate hypersensitivity reactions like LAA. Generation of these antibodies requires the central role of a type of lymphocyte known as $CD4^+$ T-helper lymphocytes.

The assessment and treatment of LAA requires a comprehensive occupational history facilitated by specifically designed questionnaires to obtain important information including the onset and severity of symptoms and correlation of the symptoms to exposures in the laboratory animal facility. Further confirmatory tests such as the detection of IgE antibodies to laboratory animal allergens (specific sensitization), skin testing to common seasonal and perennial allergens outside the workplace, and spirometry in cases of impairment of lung function are crucial.

Table 6.5.1 shows some of the major laboratory animal allergens identified and characterized. For primates, few cases of sensitivity identified with the animals' dander have been documented.

Table 6.5.1: Major animal allergens

Animal	Allergen	MW^a (kD)	Source	Biological function
Mouse (*Mus musculus*)	Mus m 1 (prealbumin)	19	Hair, dander, urine	Lipocalin-odorant binding protein
	Mus m 2	16	Hair, dander	Unknown
	Albumin		Serum	Serum protein
Rat (*Rattus norvegicus*)	Rat n 1A/Rat n 1 B (a 2_u-globulin)	16–21	Hair, dander, urine, saliva	Lipocalin-pheromone binding protein
	Albumin		Serum	Serum protein
Guinea pig (*Cavia porcellus*)	Cav p 1		Hair, dander, urine	Unknown
	Cav p 2		Hair, dander, urine	
Rabbit (*Oryctolagus cuniculus*)	Ory c 1	17	Hair, dander, urine	Unknown
	Ory c 2		Hair, dander, urine	
Cat (*Felis domesticus*)	Fel d 1	38	Hair, dander, serum	Unknown
	Albumin		Serum	Serum protein
Dog (*Canis familiaris*)	Can f 1	25	Hair, dander, saliva	Lipocalin-cysteine protease inhibitor
	Can f 2	19	Hair, dander, saliva	Lipocalin
	Albumin		Serum	Serum protein

Note: MW^a = molecular weight.

Zoonoses

Zoonosis is a disease that can be transmitted from humans to animals or animals to humans (including arthropo-zoonoses). Risk factors include the agent, host and environmental characteristics that may affect the likelihood of exposure, infection, and disease (including its severity). The disease occurs if the host is susceptible and/or the agent is present in sufficient quantity and has the necessary factors to produce disease. These agents include viruses, bacteria, fungi, protozoa and internal and external parasites.

This disease can be transmitted directly through bites, scratches, needlesticks, skin contact and mucous membrane exposure through splashes and splatters. Aerosol formation (airborne) and the faecal-oral route present the other modes of transmission. In certain cases, an intermediate host is needed for a human to be infected, such as in certain tapeworm infections. This form of disease transmission is known as indirect zoonoses. Fomites, the inanimate objects such as cages, bedding, feeding pans, scrub brushes, boots, clothing, gloves, and dust particles, provide a mechanical means of infectious disease transmission as well.

Animal carriers sometimes show signs and symptoms of the disease. However, many animal carriers do not manifest physical symptoms of the disease and can potentially transmit infectious disease to human. Hamsters, for example, show no signs of the viral disease lymphocytic choriomeningitis (LCM) when they carry the disease. Personnel who handle infected hamsters can become ill with the disease. Mice exposed to such hamsters also frequently develop serious clinical disease.

All common laboratory animals host a large range of organisms on their skin or fur, in their mouth, alimentary canal, respiratory and urogenital tracts and in their urine and other body fluids, excrement and exudates. Many of these organisms are completely or mostly harmless, but some of them are the direct causes of disease. There are also opportunistic organisms, usually harmless, which can cause diseases under certain condition. Others mutate and give rise to more virulent variants. More importantly is to note that though most infectious agents are species-specific, they can change with time in terms of virulence and ability to cross the species barrier.

The probability of contracting zoonotic diseases is very low when working with animals specially bred for research but the risk is much higher with animals caught in the wild. Nevertheless, research animals should always be treated as potential source of these diseases regardless of their origin. A general rule is that the closer the phylogenetic relationship of the animal species to man, the greater the risk for zoonoses. Hence, the use of apes as laboratory animals is often avoided and macaques are used with precautions to prevent exposure to zoonotic micro-organisms. The most notable of which is *Cercopithecine herpesvirus I* (Herpes B virus), because it can cause a fatal encephalopathy in humans. Cold-blooded animals such as turtles can also transmit certain diseases to humans such as salmonellosis, thus, proper protection should always be donned when working with these animals.

A list of some potential zoonoses is provided in Appendix 5.

Exposure Control

It is important to reduce the risk of zoonotic diseases, that is, the probability of contracting disease or exposure to infectious agents in the work environment. This can be achieved through a comprehensive occupational health and safety programme, proper education and training and appropriate risk analyses for all personnel working in the laboratory animal facility. Biosafety methods include engineering controls (for example, ventilation systems, biosafety cabinets and fume hoods and negative-pressure animal rooms), implementation of appropriate policies and standard operating procedures (SOPs), and wearing of appropriate personal protective equipment such as laboratory gowns, scrub suits, gloves and masks. Good practices and procedures also entail proper and frequent handwashing, safe and correct handling of the animals and safe handling and disposal of sharps, blood and blood products, cultures and stocks of infectious agents and contaminated animal carcasses and wastes.

The animal facility should also provide all personnel with immunizations against certain work-related infectious diseases. The most common and recommended immunizations are listed below:

- Tetanus — All individuals working with animals should be immunized with tetanus toxoid.
- Rabies — Personnel who work with random source of dogs and cats and other potential rabies carriers as well as those working with the rabies virus should receive pre-exposure rabies prophylaxis.
- Hepatitis B — Researchers and technicians who work with serum, blood, or tissues from humans or other primates should receive Hepatitis B vaccine.

Once exposure is suspected, the situation and the personnel involved need to be evaluated carefully through proper accident reporting and thorough investigation for appropriate and early treatment and considerable follow-up by an occupational health physician.

CHAPTER 7

SUPPORTING FACILITIES DESIGN

CHAPTER 7.1

CLINICAL SKILLS LABORATORY

Robert Ng

The main function of Clinical Skills Laboratory (CSL) is to serve as a venue for medical trainees to have hands-on psychomotor skills training for learning new skills and for clinicians to be upgraded and introduced to new medical technologies for better patient care and treatment. The training programme is also extended to nurses, technicians and scientists as well and is based on teaching aids and use of materials like mannequins, jigs, animals and cadavers. As such, the location of CSL should be in close proximity to animal holding area and cadaver repository that will facilitate logistics of material transfer, storage and disposal. CSL should have multifunctional capabilities in order to service the wide range of training activities covering all medical and surgical disciplines. Besides the usual surgical and veterinary courses that are being conducted, it is also the venue for medical vendors and commercial organizations to conduct training workshop courses for sales personnel and also to link up with clinical experts to showcase the application value of new developing medical technologies and devices for healthcare use.

Most of the training workshops will normally involve lecture sessions, expert demonstrations followed by supervised hands-on practices by participating trainees. The structural design of CSL will normally incorporate specialized functional unit where due considerations are given for optimized visual presentation and address biosafety issues. This chapter will focus on some of the workshop courses where experimental animals are used, which include courses in microsurgery, endoscopic surgeries, vascular surgeries and course on responsible care and use of laboratory animals.

Microsurgery Course

This five-day course (Fig 7.1.1) is so designed as to provide ample opportunities for participants to acquire the various basic techniques from adjustment of the operating microscope to the performance of a vascularized free flap surgery. The course is held in the main CSL training hall, which measure 80 × 30 feet and has 30 workstations. Each workstation comprises of a stainless steel worktable with its own dedicated power supply, sink and waste bag. Each workstation is also provided with the following training materials and supports.

1. *Operating Microscope*
   The table top operating microscope comes with a side teaching monocular side tube, two numbers of 10 × eye pieces with diopter adjustment, a magnification knob for adjustment range of up to 6 × (for suture knotting), up to 40 × (for suture placement) and focus knob for fine focusing.

2. *Instrument*
   Microinstrument set (Fig 7.1.2) comprises of two straight jeweller forceps with 0.15 to 0.30 mm tip, one angulated jeweller forcep, one spring handle curved needle holder, one spring handle curved scissors, one bremer double clamp approximator, three single vascular clips, one Adson toothed forceps (1 × 2 teeth), scalpel blade holder, two abdominal retractors that can be fashioned from paper clip and one steven scissor.

3. *Surgical Support*
   Cork board of dimension 8 inches × 12 inches for animal mounting, saline, 10 % lignocaine, gallipot, suture background material, gauze and 10/0 monofilament nylon suture.

The main demonstration worktable is located in front of the hall and equipped with the same items as the trainee's workstation except that the microscope is normally a floor model with automatic focusing and magnification function using a foot pedal. CCD camera is attached to the microscope body with outlet video cable for transmission of video images to two LCD projectors for front wall screen projection and eight numbers of 29-inch television along both sides of the room wall for viewing by those at the back of the hall. Live demonstrations are sometimes replaced by pre-recorded videotape presentation which is found to be more practical as edited video in more informative than live commentary.

Lecture is delivered from a speaker rostrum located in front and comes with connecting cabling for laptop computer presentation, which has replaced photographic slide projection as media of instruction. A built-in microphone connected to an audio mixer console is used for audio communication.

Rat is anaesthetized with intramuscular injection of ketamine/valium (50 mg/kg: 5.0 mg/kg). Two to three further anaesthetic top up with one third of the dose amount each time is required for a five-hour animal surgery. Animal is euthanized with pentobarbitone overdosing intracardially at 100 mg/kg.

For the five-day course, trainee will perform hands-on exercises on carotid artery end-to-end anastomosis, femoral artery end-to-end anastomosis, femoral vein end-to-end anastomosis, end-to-side anastomosis, sciatic nerve repair, vein graft and *in situ* groin flap transfer.

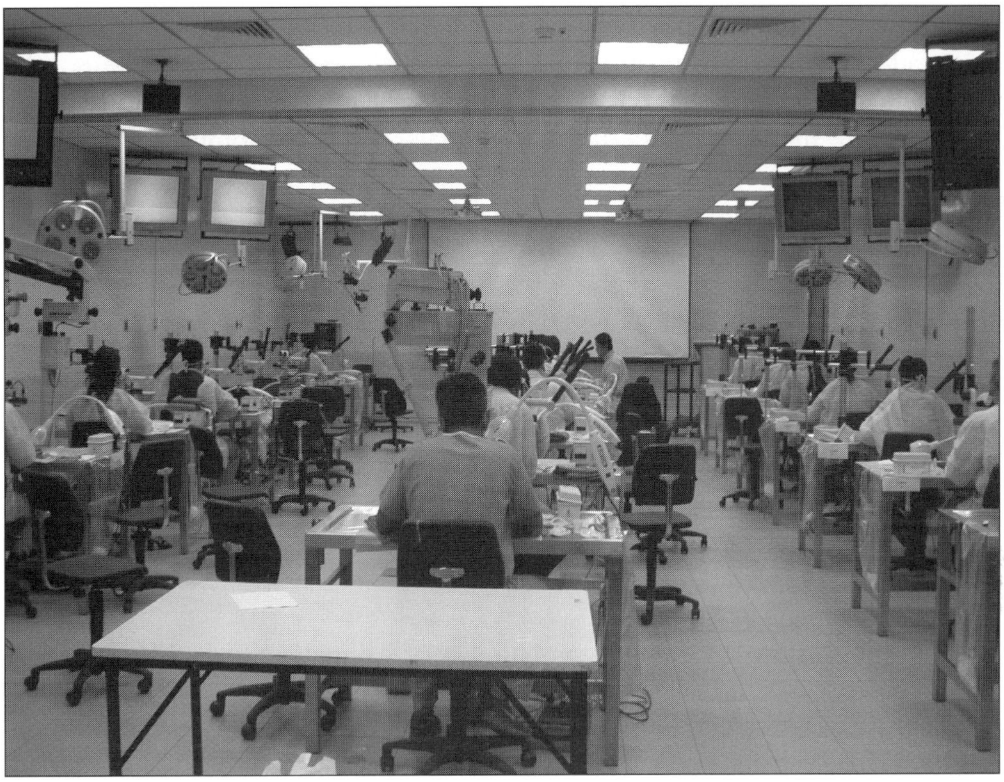

Fig 7.1.1: Microsurgery course.

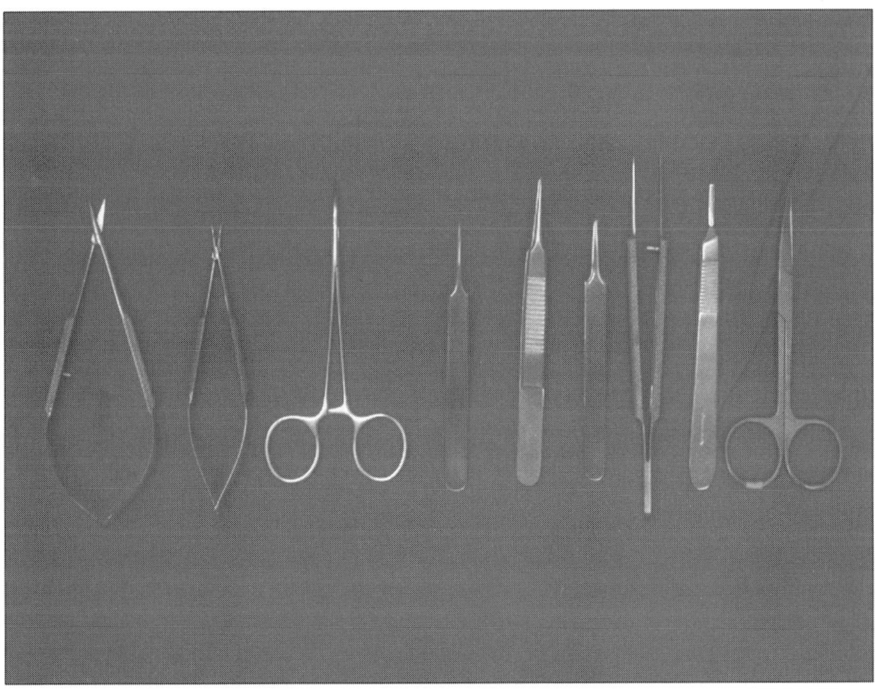

Fig 7.1.2: Microsurgery instruments.

Responsible Care and Use of Laboratory Animal Course

The course (Fig 7.1.3) is a regulatory requirement to meet the training guidelines of the National Advisory Committee for Laboratory Animal Research (NACLAR) in Singapore. The course address the prerequisite training needs of principal investigators, clinicians, scientists, medical technologists and all others involved in research procedure and training courses that require the use of experimental animals. This is a comprehensive course that will impart requisite knowledge of the legal regulations and safety guidelines and also for the participants to acquire the basic skills for handling animals in a safe and human manner. It will allow participants to develop a better understanding of the processes required in using animals for research and training. The course is a joint collaborative effort of Department of Experimental Surgery and SingHealth-IACUC and the faculty team comprised of staff and members of the two organizations. The lectures include the following topics namely,

1. Institutional Animal Care and Use Committee (IACUC)
2. Laws, Regulation and Guidelines for Biomedical Research
3. Responsibilities of Principal Investigators and Research Protocol Evaluation
4. Radiation Safety in Animal Facility
5. The 3 Rs and Research Variables
6. Occupational Health & Safety
7. Use of Statistics as Determinant for Number of Animals to be Used
8. Zoonotic Diseases in Laboratory Animals
9. Animal Anaesthesia and Pain Management
10. Animal Handling and Blood Collection.

Live demonstrations and hands-on exercises on small animal models (mice and rats) are conducted by the Institutional Veterinarian and will cover the following procedures:

1. picking up, sexing and restraining the animals
2. gavaging
3. injection Routes (subcutaneously, intraperitoneally)
4. inhalational anaesthesia
5. blood collection (tail vein, retro-orbital, tibial and maxilofascial vessel)
6. euthanasia (carbon dioxide, cervical dislocation, drug overdosing).

Two trainees are allocated to one workstation in the main CSL hall. Each participant is provided with one mouse and one rat. Anaesthesia (ketamine and valium mix), saline, syringes and gauze are provided for the animal handling practices. Rabbit is used only for demonstration of handling techniques. Full PPE is provided and floor trainers are available to assist the participants.

Fig 7.1.3: Responsible care and use of laboratory animal course.

Robotic-assisted Animal Laparoscopic Surgery Training Course

This course is conducted as a certification course in urological laparosopic surgery by the SGH Robotic Team, which have conducted over 125 robotic radial prostalectomies. The course objective is two fold. Firstly, it aims to provide interactive training in robotic-bench and "live" animal sessions and secondly to offer a comprehensive stepwise teaching programme to allow surgeons to effectively practise robotic-assisted laparoscopic surgery (Fig 7.1.4).

A pig of 30 to 40 kg in weight is used for the workshop. It is anaesthetized with ketamine (15 mg/kg) and maintained on 2 % isoflurane through inhalation. Carbon dioxide gas insufflation is provided for inflation of the intraperitoneal space via ports and maintained at 10 to 12 mmHg pressure to allow for access of scissors, needle holder or forcep which is attached to the robotic arm. The robotic arm is manipulated from a command console that is operated by thumb and index finger movement with the operator's eyes focus on an inbuilt screen for visualization of the surgical process. The stepwise workshop session will involve the following procedures:

1. draping, calibration and positioning of robotic arms and ports insertion
2. kidney dissection, ureteric dissection and mobilization for anastomosis
3. Hilar vessel mobilization and division between ligature with silk and clips
4. reposition of the animals and ports for lymph node dissection
5. dissection of fallopian tube and anastomosis to bladder (neocystectomy)
6. dissection of urethra, transection and primary anastomosis over catheter.

The initial course is targeted at urology surgeons and has since been used for training of vascular surgeons and cardiothoracic surgeons as well.

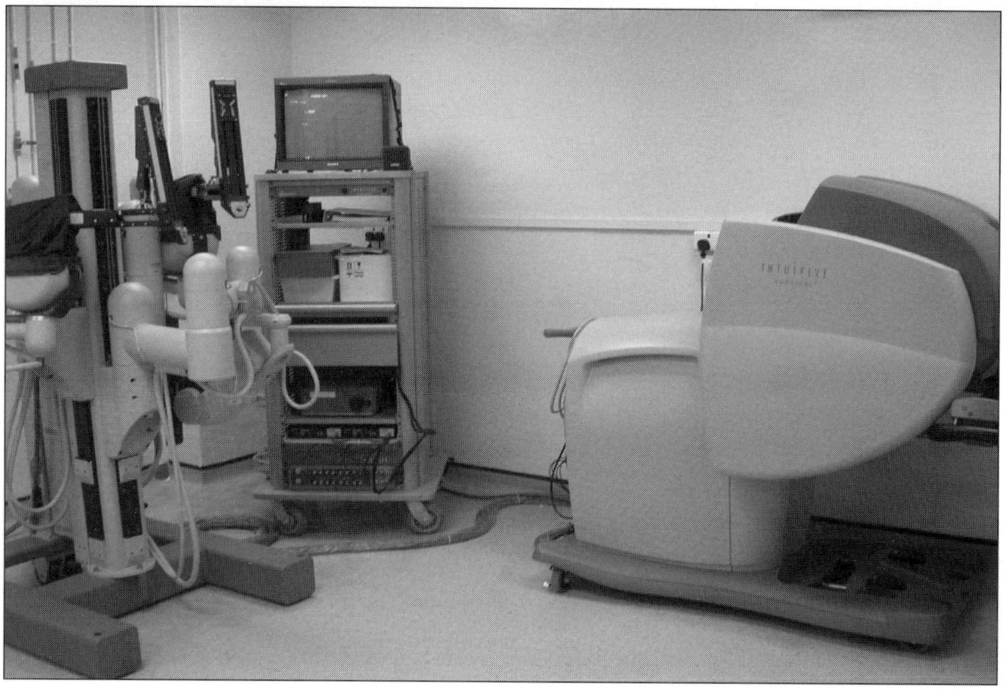

Fig 7.1.4: Robotic-assisted animal laparoscopic surgery training course.

Examples of other animal-based courses conducted in the Clinical Skills Laboratory include those listed in Table 7.1.1.

Table 7.1.1: Other animal-based courses conducted

1	General Surgery	Advanced Trauma Life Support (ATLS) Course
		Vascular Surgery Workshop
2	Respiratory Medicine	Bronchoscopic Workshop
		Arthroscopic Workshop
3	Obstetrics & Gynaecology	Laparoscopic Procedure in Hysterectomy
		Advanced Endoscopic Workshop
4	Paediatric Surgery	Advanced Stapling Workshop for Paediatric Surgical Trainees
		Paediatric Airway Course
5	Urology	Workshop on Laparoscopic Urology
		Robotic-Assisted Laparoscopic Surgery Course
6	Postgraduate Medical Institute	Basic Bowel Anaestomotic & Laparoscopic Surgery Course
		Basic Wound Closure Course
7	Experimental Surgery	Instructional Course on Basic Microsurgery
		Responsible Care & Use of Laboratory Animals
8	Cardiology and Cardiothoracic Surgery	Minimally Invasive Coronoary Artery Bypass Workshop
		Cardiac Mapping Course
9	Colorectal Surgery	Colorectal Anastomotic Workshop
10	Orthopaedic Surgery	Knee and Hip Replacement Surgery
11	Plastic and Hand Surgery	Vascular Flap and Hand Course
12	Anaesthesia	Pain Management Course

CHAPTER 7.2

ANIMAL RESEARCH SUPPORTING LABORATORIES

Robert Ng

It is preferable for an animal research centre to have not only animal holding facilities but also supporting shared laboratories, such as an operating theatre, bioimaging facilities, procedure rooms, necropsy rooms and a histopathology laboratory (with capabilities including immunohistochemistry). This will minimize the need for animal transfer and tissue outsourcing for investigative evaluations. However, due to economic reasons like high equipment cost and low volume turnover, it is logical to outsource tests such as clinical biochemistry, haematology, coagulation analysis, urinalysis and microbiology to clinical laboratories. Hence, location of the animal research centre within a university or hospital campus will be an advantage. It must be noted that equipment and assays used for human samples may not always be suitable for animal samples without some calibrations or adjustments to accommodate for species differences.

Histopathology Laboratory

Histopathology laboratory services are essential to a research facility where tissues from animal studies are to be studied. Here, tissues can be processed, sectioned and stained for histological or immunochemical evaluation of cellular changes resulting from surgical manipulations, drug administration or material implantation.

As solvents and formalin are normally used, a ducted fume chamber or overhead fume hood needs to be installed to exhaust off toxic fumes generated. There should also be provision in the laboratory for a ducted storage chamber for ventilation and waste cannisters for temporary storage of used chemical waste prior to disposal by an authorized contractor.

Since hazardous substances are used, it is important to locate a spill kit and eye wash in the near vicinity. Overhead glass shelving should have non-corrosive stainless steel hinges as a safety precaution. Equipment required and used in the laboratory will include the following listed in Table 7.2.1.

Table 7.2.1: Equipment required in an immunohistochemistry laboratory

Equipment	Use
Tissue processor	For automatic stepwise tissue dehydration and wax impregnation
Wax dispenser	For wax blocking of tissue
Freezer plate	Cooling tissue block for sectioning
Microtome	Rotary or sliding blade for sectioning
Sawing and Milling machine	Sectioning of plastic embedded tissue like undercalcified bone or tissue with metallic implants
Heating platform	For dewaxing and fixing of tissue to glass slides
Flotation bath	For floating and fishing section onto glass slides
Staining utensils	Coplin jars or autostaining system
Microscope	Evaluation of light transmitted images for staining quality
Staining tray	For immunohistochemistry staining
Incubator	For staining incubation and sectioned slides warming
Refrigerator	For delayed cooling immunohistochemical reaction

Necropsy Room

The main equipment in this room (Fig 7.2.1) is the necropsy table, a stainless steel bench top table incorporating a central sink that comes with a removable perforated top platform, on which animal necropsies are conducted. Room air can be down drafted through the perforated top by connecting the room air exhaust to the side of the central sink with capacity for 800 to 1000 cubic metre per hour (CMH) air exhausting. The down drafted air will help to protect personnel from aerosol and fume exposire during necropsies. After necropsies, remnant tissues can be deposited into double plastic bag and the bag sprayed with disinfectant (Virkon S (1:100) or other suitable disinfectant) before disposal for incineration. Loose tissues, blood or liquid waste flowed through the perforated top into the sink which drains into a bottom sealed waste treatment tank. If biohazards are involved, the waste is treated for 20 minutes with sodium hypochlorite to a final volume concentration of 1 %. The tank waste drainpiping valve is then opened for treated waste discharge into the sewer.

A carbon dioxide chamber is available for rodent euthanasia and is comprised of a perspex box with a covered top that has a gas port for delivery of carbon dioxide gas from an inhouse wall mounted gas delivery system. An operating microscope is provided to facilitate necropsy procedures if required. A standard set of surgical instrument is also provided which include scalpel blade holder, dissecting forceps, tissue forceps (rat tooth), self retaining retractor, metzembaun scissors, steven scissors, mosquito artery and jeweller forceps.

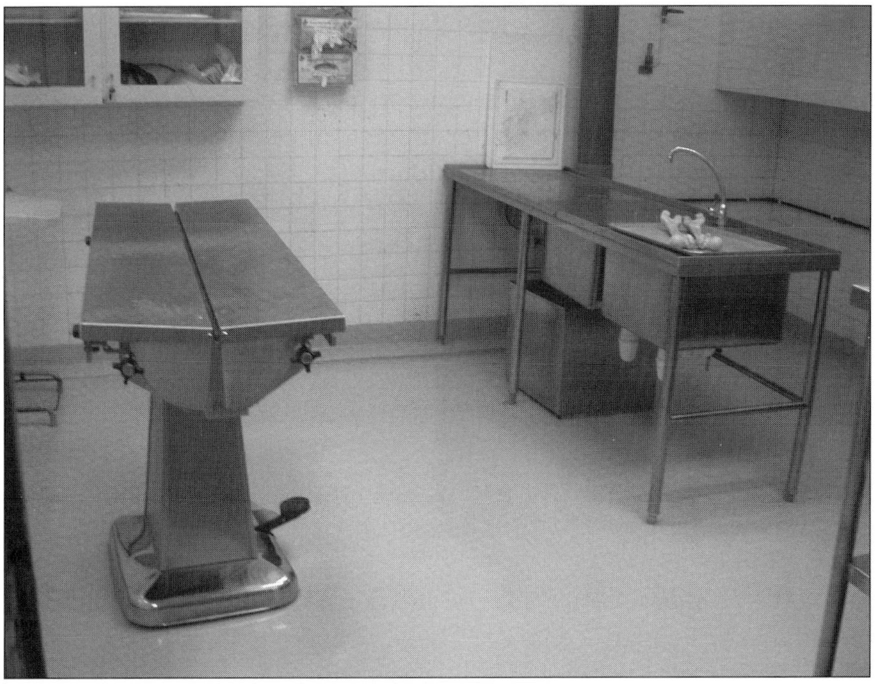

Fig 7.2.1: Necropsy room.

Quarantine Station

Animal quarantine stations are required for temporary holding of animals that are delivered from unspecific sources or when full health certifications for the animals are not furnished. The size of the quarantine station is dependent on the frequency and volume of animal usage. For a small establishment, a normal rodent quarantine room size of 12 by 20 feet should suffice and will accommodate three or more self-contained, individualized cubicles located along side of the room wall. Since all room air exhaust ducts are located in the cubicles, air supplied to the room will flow unidirectionally through grills into each cubicle providing 15 to 20 air changes/hour ventilation. Static micro-isolator cages (shoe boxes) on stainless steel racks will maintain isolated micro-environments. Animals are monitored daily and blood test performed to verify health status before being allowed into the main holding area. It is important that air supply and exhaust controls are connected to an emergency power supply to ensure uninterrupted airflow and reduce chances of possible contamination of other rooms due to air backflow during power failure. Seamless ceiling of bonded calcium silicate board or concrete is preferred. Walls and floors are painted with epoxy paint with cement or silicone shirting along floor and wall junction. All surfaces in the room must be impervious to moisture and fully sanitizable.

Procedure Room

It is ideal to have separate procedure room for each animal species to minimize potential cross contamination between species. It is also justifiable to maintain a sterile environment to facilitate conduct of animal procedures in clean room set up with HEPA-filtered 100 % fresh air supply to a positively pressured room. A Biosafety Cabinet Class II will be an added biosafety advantage. Normal disposables like gauze, syringes and tubes are provided. Heated bead sterilizer is used to expedite instrument sterilization for minor surgical procedures. A wall mounted oxygen delivery port provides the anaesthetic gas inhalation for the animal with isoflurane via vapourizer unit. A ducted or activated charcoal cannister scavenging system is required for waste gas handling.

Bioimaging Centre

The room facility set up (Fig 7.2.2) is similar to that for an animal procedure room. However, because radioactive tracers are used there is a need to have lead blocks to screen personnel from the gamma ray emission effects from used vials and animals injected with radioactive substances. Where emission is from X-ray or C-Arm fluoroscope lead lining of wall is mandatory to be in compliance with Radiation Protection Inspectorate. Where beta ray emitting substances are used the installation of 1 cm thick perspex or lucite screen is necessary for screening.

Normally short-life radioisotope use is advocated and use of luminescence as an alternative should be considered. A Grenier counter should always be available to detect for leaks or spillage. Personnel working within the facilities must wear a dosimeter badge, which is sent to the Centre for Radiation Protection for monthly safety dose exposure level. Lead aprons must be worn by all personnel working with X-ray or Fluoroscope.

Piped in oxygen source is provided for animal anaesthesia with isoflurane inhalation with a gas vapourizer. An anaesthetic waste gas is scavenged off via a ducted wall mounted scavenging port or an activated charcoal cannister.

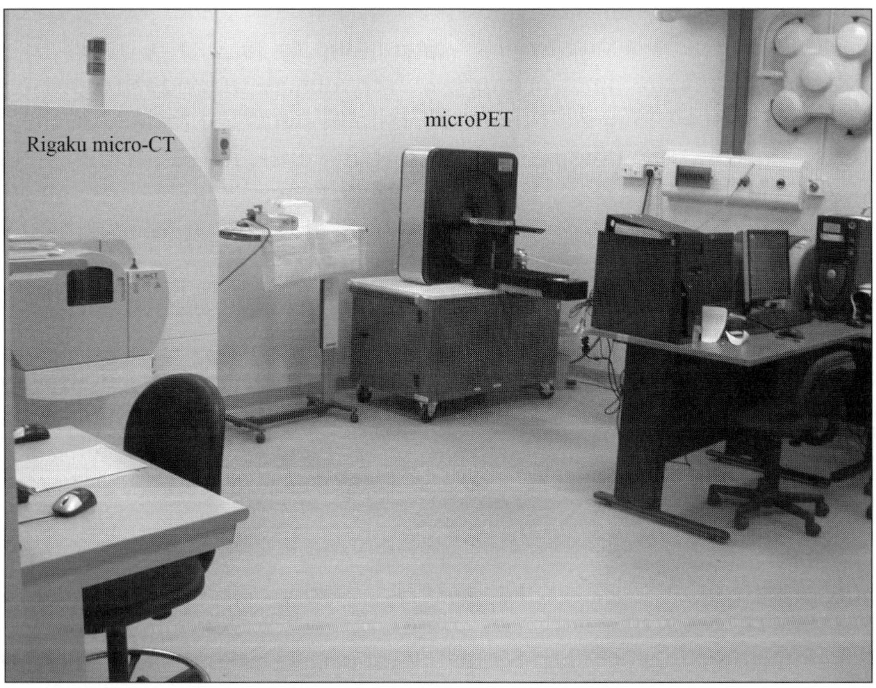

Fig 7.2.2: Bioimaging room/centre.

Operating Theatre

The Operating Theatre (Fig 7.2.3) caters for surgical procedures on big animal models like rabbit, monkey, pigs and sheep. Each animal surgical workstation comes with the following equipment (Table 7.2.2).

Table 7.2.2: Equipment provided in each animal surgical workstation

Equipment	Description
Operating Table	Has hydraulic pedal for vertical translation and turnwheel for lateral and front to back tilting. It also comes with in built heated table platform, or can be topped up with an electrically heated mattress.
Operating Light	Cool light delivery from 3 to 5 lamps configuration providing 80,000 to 100,000 lux illumination for a lighted field of diameter 30 cm.
Anaesthetic Machine	Provision should be provided for 3 operating modes for manual, spontaneous and Intermittent Positive Pressure Ventilation (IPPV). Adjustable control of tidal volume, delivery pressure, respiratory rate and inspiration/expiration ratio. Incoming medical oxygen and air are supplied from wall mounted piped in gas port. A scavenging port is also provided for evacuation of waste anaesthetic gas from the machine.
Physiologic Monitor	Used to register the animal patient vital signs such as ECG, respiratory rate, pulse oximetry, artery pressure, heart rate and temperature. End-tidal CO_2 measurement is also desirable.
Diathermy Machine	Used for monopolar and bipolar coagulation and cutting of tissue.
Suction Machine	There is a tube connection to suction probe for evacuation of blood leaks or lavage fluide from operative field by vacuum.
Drip Stand	This stand is used for attachment of fluid bag to intravenous infusion.
Pressure Transducer	This is a dual connection by tubing to pressurized fluid bag and arterial cannula for transfer of pulse signal via electrical cable to physiologic monitor for blood pressure registration.
Mayo Stand	For display of sterile surgical instruments within easy research of the surgeon.

Sterility of Operating Theatre is a mandatory requirement. Room ventilation is provided by HEPA-filtered 100 % fresh air supply with calibrated sizing of the exhaust ducting to generate room positive pressure and maintenance of air circulation for 15 to 20 air change per hour. As further enhancement of the clean room environment an annual particle count is conducted for compliance to IS Class 8 status.

Floor of the OT should have a smooth surface with either epoxy resin application or vinyl sheet surfacing for easy sanitation. Walls are to be painted with epoxy coating and ceiling should be of non-gridded, non-porous structure such as concrete or bonded calcium silicate boardings for a smooth surface with epoxy paint. Adjoining the OT is a change room for personnel to change from street wear to blue scrub and access to the OT is via a Scrub Room for hand scrubbing and sterile gowning. A separate entrance is designated for animal access from the surgical preparation area.

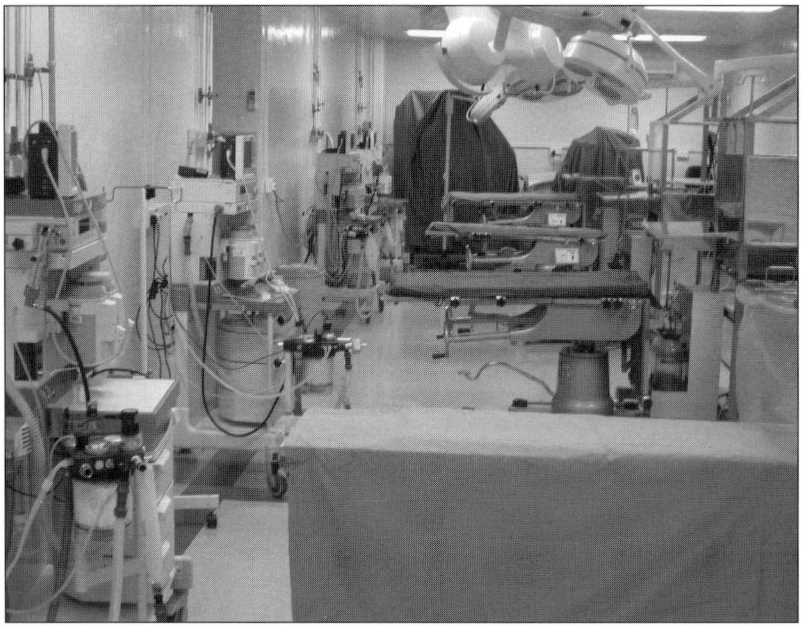

Fig 7.2.3: Major operating theatre.

Microsurgery Laboratory

The Laboratory (Fig 7.2.4) is similar in structure to the Operating Theatre and is designed for rodent surgeries. Each surgical workstation consists of an operating microscope, microsurgery instrument set, inhalation anaesthesia chamber and glass bead sterilizer. Inhouse piped in medical air and oxygen is delivered via wall mounted gas port through a vapourizer. A mini membrane pump ventilator is also provided in cases IPPV is required.

Fig 7.2.4: Microsurgery laboratory.

CHAPTER 7.3

ANIMAL RESEARCH AND HOUSING FACILITIES

Robert Ng

An animal research centre set up should make provision for segregation of functional units, namely administration, laboratories and animal handling areas. This will facilitate implementation of biosafety measures and design of structural layout that will best reflect the operational needs of each segregated component. This can be achieved if the facilities are located at different building level. For animal research centre that is located at a single building level, there is the need to install buffer zone either in the form of anterooms, possibly with air locks or void space demarcation.

Environment variables in an animal research centre may have direct impact on biosafety controls, which may compromise occupational, health and safety compliance. Such variables may also impact animal health status, which may affect the outcome of a research study. These variables, which occur in both the micro and macro environment, must therefore be stringently controlled. Considerations should be given to environment parameters, which include room ventilation, temperature and humidity, light intensity, sanitation, vermin control, clean and dirty area definition and hazardous material usage and disposal.

Room Ventilation

Air conditioning systems are normally used to cool the air for delivery to the rooms and they come in different configuration to meet suit specific uses. Recirculating systems can be used for administrative offices, conference room and store rooms. A ceiling or wall mounted fan coil air conditioner unit will supply 70 to 80% recirculating air delivery. Animal housing rooms must have 100 % fresh air supply. The air handling unit, usually centralized, cools the air supply through chilled water coils before delivery through above ceiling ducting and diffusers for distribution into rooms. High Efficiency Particulate Arresting Filters (HEPA filters) can be used to cleanse the air for clean rooms. Clean room environments are often indicated for the Operating Theatre, Microsurgery Laboratory, Animal Procedure Room, Tissue Culture Lab, Bioimaging Centre, BSL Facility and some SPF Animal Holding Area. For some *in vitro* laboratories there may not be need for clean air supply and HEPA filtration is unnecessary. For animal surgical facilities, like the Operating Theatre and Microsurgery Laboratory, an annual air borne particle test for clean room certification for ISO Class 8 (ISO14644-1: 1999E) is conducted under at rest condition will be useful.

Air Change and Pressure

Sizing of the exhaust fan and control of air supply by adjusting the supply duct damper mechanism will regulate the volume and flow, measured in m^3/hour, to generate the desired air change requirement for the room. Ventilation for animal housing rooms is normally set at 15 to 20 air change per hour. Air volume and flow can be measured with an electronic balometer.

Regulating the air supply volume and exhaust discharge rate will also establish the room air pressure relative to external spaces like the corridor or adjoining areas. Since air flows from high to low pressure source, this concept is used to generate containment areas to prevent contaminants from coming into or flowing out of the room according to room needs. Room facilities, like necropsy room, quarantine room and conventional animal holding facilities including for nonhuman primate housing, are kept at negative pressure. This is necessary to prevent contamination of clean animal colonies and to protect personnel from animals that may pose as biohazard.

BSL 2 facilities must be maintained with negative pressure, have anterooms, a biosafety cabinet and an accessible autoclave. BSL 3 facilities are specialized and complex structures, where high differential air flow in air tight rooms are required to generate negative pressure of –15 to –30 Pascal, and may incorporate features like bag in bag out HEPA filters, double door autoclave, decontamination shower and air lock anteroom. Conversely, sterile areas like the Operating Theatre, SPF Animal Holding Area, Microsurgery Laboratory, Tissue Culture Laboratory and Bioimaging Centre are maintained at positive pressure to the external area to prevent contaminants from entering into the rooms.

The microenvironment in the rodent isolator cage has its own dedicated air supply and exhaust ports connected to a ventilation console that allow for adjustment of negative or positive pressure and also reduce ammonia and carbon dioxide levels inside the cage.

Temperature and Humidity

Humidity and temperature are two important variables, which should be recorded and monitored daily. Temperature can be measured with wall mounted maximum and minimum mercury bulb thermometers and humidity with a dial hydrometer. A combined thermo/hydrometer with memory plug-in cable sensor and LCD display that can be mounted outside the room may offer a better option as it is wireless and does not use mercury. Although acceptable temperature range may be 19 to 26 °C and humidity range of 50 to

70 %, it is important to ensure that daily temperature fluctuation must be within a narrow bandwidth of 3 to 4 °C, customized for each species. Extremes in relative humidity and wide daily variance in temperature should be avoided as the animal can be stressed and predisposed to diseases.

Power and Lighting

The electrical power system should be safe and sufficient to satisfy the required number of power outlets in the room. Although most power points are installed for 13 amp three-pin plugs, suitable amperage should be provided for specialized equipment like 15 amps for freezers, 3 phase 20 amps for normal autoclave and 3 phase 50 amp for tunnel cage washer and large autoclave. Critical equipment like freezer, refrigerator and ventilated caging system should be powered from emergency power points with backup power supply from a standby generator, that will be activated in seconds when there is power failure. Wall mounted power points should be situated at least one meter away from water sources and lighting points that are susceptible to water contact should have plastic cover shields.

Light intensity in the room should be in the range of 300 to 350 lux measured at one meter above the floor. Lighting that is too intense has been shown to cause retinal degeneration in the eyes of albino species. For enclosed animal room, time controlled system is required to ensure uniform 12 hours light/dark daily lighting cycle for most species. Variation in light cycle can result in reduced breeding in some species, especially in rodents.

Noise Control

Noise level in the animal room should be kept to a minimum and sudden loud noise greater than 80 decibels (db) from cleaning machine; radio or phones should be avoided or prohibited. Facilities that generate noise during operational activities like cage washing (80 to 100 db), should be sited away from animal housing and experimental function areas or should have sound proof partitioning or creation of buffer zone separation. Masonry walls are effective barriers against noise transmission. Calcium silicate partition walls should have double wallpanelling encasing insulating fibre to screen off sound passage. Some small animal species like rodents are able to detect high frequency sound hence equipment that are capable of generating ultrasonic frequency sound should be located away from these animals. Some rodents are susceptible to audiogenic seizures or tetranic-like spasms, increased blood pressure and auditory damage when they are exposed to sudden or sustained loud noises.

Storage Areas

Corridors are designed for passage of personnel and for fire escape routes and should therefore not be used for storage. Feed and bedding stores are designed for cool environment (less than 21 °C) and low humidity (less than 50 %) to minimize deterioration. Feed and bedding bag are stored on rack shelves or plastic pallets with the lowest shelf at least 9 inches above the ground. The stacked feed bags should be kept at least 6 inches away from wall surfaces to minimize access by ants and other pests and to prevent contact with moisture that may condense on the walls. Mousetraps or sticky bait are provided for detection and trapping of vermin. Doors should incorporate auto hinged closure mechanism and false ceiling should be completely boarded with no exposed gaps for vermin entry. Plastic container with cover is used for storage of unused or open bags of food. New and old feed supply should be rotated to ensure close adherence to expiry dateline of about 6 months from manufacture date provided that stabilised form of Vitamin C is used as an ingredient. Irradiated diet may be considered as an alternative to autoclave feed as sterilization can result in uneven nutrient loss and hardening of the feed pellets.

Designated waste store should preferably be sited at a location that allows for convenient exit to outside the facility for disposal. Biohazard waste bins are normally provided by waste handling contractor and are collected for disposal on the same day. Some centres have refrigerator spaces (cold rooms) for temporary storage of animal carcasses prior to disposal for incinerations. Carcasses should be double bagged and spray disinfected before disposal as biohazard material. A designated refrigerator is also provided for temporary storage of radioactive waste like short half-life radioisotopes or nuclides for degradation and measured with a Grenier Counter before they are disposed. Lead shielding (for gamma radiation) and 1" thick perspex shielding (for beta radiation) are provided together will appropriate signages.

Room Interior Construct

Animal room should be constructed in a manner that allows for easy sanitation, noise proof and durability. Walls and floor should be smooth, moisture resistance, nonabsorbent and free of cracks. Wall finishing with epoxy paint and floor surfacing with epoxy aggregates should satisfy the required criteria. Wall and floor junction should also be skirted with solid coving and noncementitious screed or silicone sealant to seal off gaps especially when wall and floor contruct are of different building material. It may not be advisable to have smooth epoxy flooring for animal pens as hoofed animals like pigs and sheep tend to slip on them. Noncementitious sealant floor surface would enhance traction property especially for postoperative recovery, to facilitate animal ambulatory activities.

Drains may not be necessary for rodent rooms as they can be sanitized satisfactorily by wet vacuuming or mopping. However, for animals housed in meshed caging system or in animal pens, scupper drains of 4" to 6" width should be provided on either sides of the room and floor gradient should be so constructed for easy drainage flow to the drains that discharge into floor traps. Waste drainpipes discharging into sewer should have elbow joint for water trap to keep out backflow of sewer gases.

Ceiling formed by concrete floor above are satisfactory provided they are smoothed and painted. Suspended ceiling with grided metal bars to hold acoustic plasterboard are normally unacceptable. Ceiling must be of seamless bonded, calcium silicate board and surface smoothened with sealing cement and finished with epoxy paint. Exposed pumbling, ducting and light fixtures are undesirable unless the surfaces can be easily cleaned. Doors should opened into animal rooms and installation of viewing window should be considered for safety purposes. Doors should be wide enough for easy passage of equipment and to fit tightly within their frames to keep out vermins. A door sweep is usually necessary on the bottom of doors. For wide access doors, it might be useful to consider double leaf door contruct.

Corridor

Corridors should be wide enough to facilitate movement of personnel and equipment and are normally about six to eight feet wide. Floor wall junction should be covered up with solid coving and sealant for easy cleaning. Corridor should have recessed spaces for the siting of fire alarms, first aid box, emergency shower, telephone and spill kit so as not to hinder movement of large equipment. A dirty-clean corridor system is preferred when space allows. The corridor along the animal rooms leading to the cage wash area is defined into dirty and clean corridor. Dirty corridors are designated for transport of soiled items while clean corridors are for traffic flow of clean items. This demarcation is structured for single directional flow of traffic along the designated corridor to prevent cross contamination and enhance safety.

Sanitation

Design of an animal facility should incorporate features that will facilitate operation of good sanitation practices. Sanitation does not only refer to the cleaning of room surfaces or corridors, but also involves cages, animal pens and related equipment to offer a total solution to a healthy environment for animal well-being.

Selection of disinfectant agent for animal facility sanitation should exclude those designed to mask animal odours as they can expose animals to volatile compounds that might alter their physiologic and metabolic process. Normal chemical disinfectants used include quaternary ammoniums, chlorine compounds and peroxidate compounds with surfactant.

Soiled bedding in rodent isolator cages are usually changed weekly or fortnightly but may vary depending on the number of animals housed per cage, urine and faecal output or when research objective does not permit changing of bedding. Disinfecting isolator cages will involve removal of soiled bedding, washing with a disinfectant soap, and rinsing in a tunnel cage washer at 84 °C.

For animal pens, daily flushing with water and periodic use of disinfectants are usually appropriate to maintain clean and sanitized pens. Rabbit, guinea pig and hamster produce urine with high concentrations of protein and minerals, which will adhere to cage bottoms and sanitary tray surfaces, hence these cages should be pretreated with perchloric or phosphoric acid before washing. Drinking bottles are cleaned with high-pressure sprayer manifold while sipper tubes and other small equipment can be disinfected with chemical in an ultrasonic bath.

Sanitation of necropsy tables, after procedure, involving biohazards, will involve collection of waste discharge in a treatment tank located at the base of the table and disinfecting with sodium hypochlorite to a final concentration of 1 % for 20 minutes before discharging into the sewer. Cages housing SPF rodents are required to be sterilized together with the bedding at 121 °C for 30 minutes in an autoclave.

Dirty carrier trolleys for soiled cage transport will be required to be fogged with hydrogen peroxide and spray clean in an adjoining room before transferring to the clean cage store area for reuse. Rooms are sanitized with chemical disinfectants using mops for the room interior. The mop head is changed daily after use and a different mop is dedicated for each room housing animals of the same species or SPF status to prevent cross contamination. Corridors may be disinfected using a scrubbing and wet-vacuuming machine.

Sanitation efficiency should be regularly checked. Autoclaves are checked with 3 M steri-strip and/or spore ampoules. Cages and room surface sanitation is verified with RODAC plates for microbial colony count. The rodent sentinel evaluation programme also helps validate sanitation efficiency.

CHAPTER 8

THE DEVELOPMENT OF COMPREHENSIVE ANIMAL FACILITIES IN SINGAPORE

CHAPTER 8.1

HISTORY OF THE DEPARTMENT OF EXPERIMENTAL SURGERY AS A REFLECTION OF TRANSLATIONAL RESEARCH DEVELOPMENT IN SINGAPORE

Robert Ng

The Department of Experimental Surgery (DES) was started in 1982 as a small research unit in the Singapore General Hospital (SGH). It has over the years developed and grown in tandem with biomedical research developments in Singapore. This chapter hopes to capture this development through the path that DES took in its history as it transformed itself from its early beginnings to maturity as a centre for translational research.

The founding committee for the Experimental Surgery Unit (ESU) was formed in 1981 and was committed to improving the training of young surgeons and saw the importance of introducing and encouraging basic biomedical research into the advanced surgical training programme in Singapore. The chairman was Dr Chew Chin Hin, then Deputy Director of Medical Services (Hospitals). The committee had, as its members, Dr Moses Yu, then Assistant Director of Medical Services (Support Services) as Vice Chairman; Dr Jimmy Sng Ewe Hin, Director of Pathology; Professor Robert W H Pho, from the University of Singapore as the Unit Coordinator; Dr Wong Kum Leng, Medical Director of Singapore General Hospital; and Dr Gopal Baratham, Head of Neurosurgery II, Tan Tock Seng Hospital as committee members. The committee's main responsibility was to develop the Unit into a centre for young surgeons to be exposed to and to enhance their skills through experimental surgery, laboratory-based research and laboratory-type training of surgical skills. The Experimental Surgery Unit (ESU) was officially set up on 12 May 1982 through the endorsement of Dr Andrew Chew, Permanent Secretary (Health)/Director of Medical Services.

As coordinator of ESU, Prof Robert W H Pho, an Orthopedic surgeon, was responsible for the day to day running, recruiting of manpower, raising of funds, planning and implementation of training programmes and research directions. This was only the beginning - the space was given but with no equipment or manpower. The resourcefulness of the committee brought them through several avenues and to different sources, obtaining excess hospital equipment, some condemned equipment, an operating lamp here, an operating table there, and a reconditioned anaesthetic machine. Everything counted then. The next priority was for manpower development. Initially ESU was given part time staff and one retired staff. Then came the secondment of Mr Robert Ng and later Ms Song In Chin from the Pathology Department to oversee the day to day running. This was a big step forward but another problem was encountered - the lack of interest on the part of the clinicians due to their heavy service commitments and the fact that little recognition was given to research and publication in experimental work.

In spite of this drawback, with strong determination from the committee members the Unit surged ahead, progressed and grew, organizing training programmes, cadaver exercises and seminars for surgeon and trainees. In fact, the progress was immediately assessed (within six months) by the then Minister of Health, Mr Howe Yoon Chong (Fig. 8.1.1) during a visit in November 1982. This was a significant visit as it immediately emphasized the potential of experimental research and its importance to advancing medical service in this country. The position of ESU was strengthened further by a second Ministerial visit three years later in 1985 by the then Minister of Health, Dr Richard Hu (Fig. 8.1.2) and Mr Yeo Cheow Tong, then Minister of State for Health. Several national firsts were also achieved at ESU: the first Seminar in Experimental Surgery from 4-6 April 1986; the first Flap Dissection Workshop in November 1986 and the first experimental liver transplant done in 1983. A research culture seemed to have finally developed.

By 1st April 1989, with the restructuring of the Hospital, and the departure of the University Departments to their new campus in Kent Ridge, a change in leadership was necessary. A/Prof Tan Ser Kiat, an Orthopedic Surgeon, took over the appointment as the new Director of ESU. The administration of ESU was at the same time transferred from the Ministry of Health to the restructured SGH. Simultaneously the Unit was conferred the status of a Department - the Department of Experimental Surgery (DES) - under the restructured SGH and came under the jurisdiction of the Division of Surgery.

Under the new management, A/Prof Tan Ser Kiat adopted an 'open-door' policy in making the Department of Experimental Surgery (DES) facilities available to all medical investigators interested in surgical laboratory research. The aim was to expand the scope of activities in DES, placing emphasis on two major areas - Basic Research in Surgery and Psychomotor Skills Development. DES was also fortunate to receive the support of a number of funding bodies without which it would have been extremely difficult to achieve its goals. Funding in the form of research grants, donations and other non-monetary support came from the Lee Foundation, the Shaw Foundation and the National Medical Research Council. Many young surgeons and doctors from within SGH and overseas were able to conduct a wide variety of research. A list of awards received by the investigators bears testimony to the efforts put into research activities by the Department. The operation of laparoscopic cholecystectomy, which took the world by storm when developed in the late 1980s, was introduced into Singapore in February 1990. The success and attractiveness of the operation generated a sudden explosive rise in demand for laboratory training and DES becomes an extremely busy training venue and in the process gained recognition for the facilities it provided for the training of surgical skills.

Fig 8.1.1: Visit by the then Minister of Health, Mr Howe Yoon Chong, 1982.

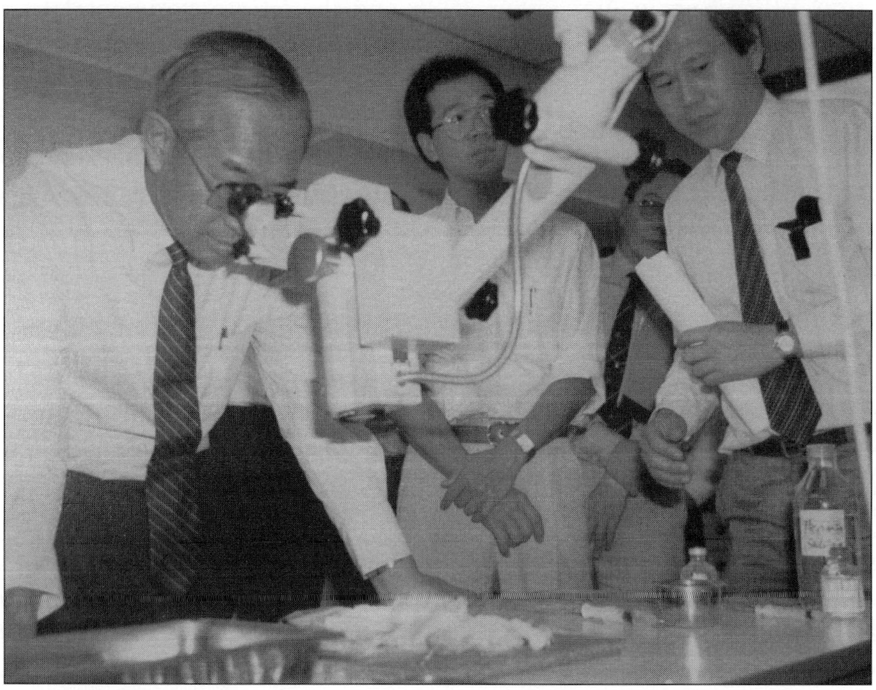

Fig 8.1.2: Visit by the then Minister of Health, Dr Richard Hu, 1985.

In May 1993, Dr Peter Mack, a General Surgeon, was appointed Deputy Director. On 1 January 1994, A/Prof Tan Ser Kiat passed the baton of Directorship to Dr Peter Mack for him to take experimental surgery research in SGH onto the next stage of development. A 5-year concept plan was drawn up to include an immediate upgrading of the infrastructure, active acquisition of basic research facilities, soliciting of funds, recruitment of scientific staff and finally collaborative ties with overseas research centres. The Practical Hall in Department of Pathology was refurnished on 3 February 1994 and re-named Clinical Skills Laboratory in line with a similar concept launched by the Royal College of Physicians and Surgeons of Glasgow at around this time. In November 1994, DES published its first book entitled "Clinician Guide To Experimental Surgery" as a reference for aspiring clinicians who are keen to conduct experimental surgery research. As more clinicians and doctors came to use the department facilities for their research projects, funding was approved by SGH-Finance for the service of Ms Irene Kee Hwee Cheng who joined DES on 2 April 1995 as a Laboratory Technician to help ease the workload. In August 1995, a manpower proposal for scientific officers was raised and supported by SGH Medical Board, which redirected the proposal to National Medical Research Council (NMRC). The proposal was accepted by NMRC and an Institutional Block Grant (IBG) was approved beginning from FY 1996. For the first time in history, DES was able to employ two scientific officers, Drs Yang Er Bin and Zhang Kai. In 1997, the Department established ties with Wallenburg Laboratory of Malmo University Hospital, Sweden and Guangxi Medical University, China as part of our initial effort to established overseas collaboration to broaden the research scope of DES.

Dr Pierce Chow, a General Surgeon, was appointed Deputy Director on 1 July 2000 and become Director on 1 July 2001. During the period 2000-2006, tremendous changes took place that transformed and prepared DES to be functionally equipped to service translational research activities. These changes coincided with the beginning of the Biomedical initiative in Singapore and were in many ways a response to increased demand for translational research services by researchers. On August 2000 at a meeting initiated by Prof Lim Yean Leng, Head, National Heart Centre and Chair NMRC with Dr Peter Mack and Dr Pierce Chow, the idea was mooted to establish an off-site Animal Husbandry & Hospital to address the need for postoperative convalescence of large animals. The Sembawang Research Station of the Agri-Food & Veterinary Authority (AVA) was then available and was identified as a suitable site. A proposal to convert the Sembawang Station into a biomedical research facility was submitted to BMRC and in-principle approval was subsequently given by Mr Philip Yeo, Chairman of EDB. On 31 July 2001, Prof Louis Lim, Executive Director, BMRC approved a sum of \$3,378,000.00 for equipment and 5-year manpower support towards the development of DES with \$670,000.00 being assigned for the renovation of an Animal Husbandry & Hospital (AH & H) facility in Sembawang. The lease for the land parcel in Sembawang Research Station was signed on 1 December 2001 as a TOL lease agreement.

At the same time plans were also drawn up to increase the approximately 1000 sq. ft existing research facilities at Block 9 level 2 to more than 2000 sq. ft by expanding into level 3, the existing rooftop. The Endowment Fund of DES was committed towards this project. On 21 March 2002, after a series of meetings, A/Prof Donald Tan of SERI approved, with the endorsement of the Board of the Singapore Eye Research Institute (SERI), that \$964,000.00 from a SERI Block Grant from NMRC which remained un-utilized, be used to support the efforts of DES in this direction. The new facilities at Block 9 level 3 comprised structures for rodent and nonhuman primate facilities. The structural make-over of DES at the Outram campus generated a comprehensive translational research facility within a single institution that included:

- rodent and nonhuman primate centres in Block 9 Level 3;
- large animal facilities in Block 9 Level 2;
- supporting research facilities: radiology, biochemistry, molecular biology, cell and tissue-culture, histopathology, micro-vascular surgery
- Clinical Skills Laboratory for hands-on skills training;
- Offsite Animal Husbandry & Hospital in Sembawang for animal convalescence and the breeding of large SPF animals which could not be easily purchased;
- a Cadaver Repository in the Outram campus for anatomical research and training.

Concomitant with infrastructure development, Dr Pierce Chow envisioned the need for DES to embark on an accreditation process to position DES as a premier animal research centre meeting the best of international standards so that research conducted at DES will be widely accepted. A decision was made for DES to work towards accreditation with the Association for Assessment and Accreditation of Laboratory Animal Care (AAALAC) International. Attaining this will demand significant commitment and resources on the part of the institution and a series of meetings were held with administration towards this goal. While this initially met with skepticism and resistance, positive developments in SGH administration together with the government-sponsored biomedical initiatives lead to support for the accreditation process which was then expedited. To spearhead the process, Dr Paul Gamboa Pineda DVM was employed as an Assistant Veterinarian on 31 January 2002 to oversee and implement compliance requirements.

Historical events occurred at the same time in the Outram campus. The Division of Research in SGH was formed on 2 March 2001 under the chairmanship of Prof Woo Keng Thye and DES was transferred to this Division. The SingHealth Office of Research was set up in 2002 to consolidate and coordinate research activities and development. One of the requirements for AAALAC accreditation was the need for an Institution Animal Care and Use Committee (IACUC). The first IACUC in Singapore was thus formed on 15 October 2002 and comprised members appointed by the then CEO, Prof Ong Yong Yau. SingHealth organised the first IACUC training course in Singapore, which was conducted by Drs Ron Barne and Molly Greene of PRIM & R, USA on the 26 and 27 October 2003. The Animal Care and Use Programme required by AAALAC was submitted by Dr Paul Gamboa Pineda with SingHealth providing the funding support for a Programme Status Evaluation (PSE) visit, which was conducted by Drs Kathryn A. Bayne and Ronald M. McLaughlin of AAALAC International on December 2003. Dr Paul Gamboa Pineda resigned on 30 June 2004 and Dr Bryan Ogden DVM, ACLAM was engaged as Institutional Veterinarian on 1 September 2004 on a part-time basis. He was subsequently joined by 3 assistant vets.

On the recommendation of the SingHealth Scientific Director, Prof Malcolm Paterson, the IACUC in SGH was expanded to become the SingHealth-IACUC and membership was increased from 6 to 12 with new members being appointed by SGH CEO, Prof Tan Ser Kiat. Under the term of reference, the SingHealth-IACUC has oversight of animal facilities both in DES and National Cancer Centre (NCC) and provides scientific review of research and training proposal. It also reviews the institution's animal care and use programme every 6 monthly and conducts site inspection annually.

The National Advisory Committee for Laboratory Animal Research (NACLAR) was formed in 2003 and Assoc Prof Pierce Chow was invited to be a member. NACLAR Guidelines on the Care and Use of Animals for Scientific Purposes was finalised and announced on 29 October 2004. This subsequently became legislated and a license to run an animal research facility became mandatory for the first time. AVA became the official enforcement body to ensure compliance with NACLAR guidelines, conduct site inspection and program review of all animal research facilities in Singapore. As a result of ongoing preparation for AAALAC accreditation, DES was well prepared and was inspected on 16 and 18 March 2005 and was subsequently granted full license. NACLAR guidelines require that researcher using animals for research or clinicians conducting training courses using animal models attend a course on the Responsible Care and Use of Laboratory Animals prior to handling animals. DES organized the first course on the Outram campus for researchers on 8 May 2004 at the Clinical Skills Laboratory, which was newly refurnished with renovation cost sponsored by Zimmer.

As a final preparation for accreditation, two committees were formed — the Animal Facilities BioSafety Committee chaired by Dr Fong Yoke Tien of SGH-Occupational Health & Epidemiology and the Emergency Crisis Management Committee headed by Dr Bryan Ogden. A Programme Description of Animal Care and Use was submitted on 1 December 2005 and DES was audited on 9 and 10 March 2006 by the AAALAC team, which comprised of Drs Kathy Laber and John Bradfield. On 22 June 2006 AAALAC International informed the hospital that DES has been awarded full accreditation, making it the first institution in Singapore to be accredited and one of the first in the region.

In response to new research technology and changing trends in the use of animal models, DES brought in the first micro-PET scan to Singapore in late 2005 through a competitive research grant from NMRC topped up by SingHealth. Personnel were sent to the United States for hands-on training in the use of the machine and the endeavor was strongly supported by the Dept of Nuclear Medicine in SGH. The following year, DES brought in the first micro-CT machine to Singapore as a joint collaboration with J Morita of Japan and this was followed in 2007 by a Bio-luminance machine by Duke-NUS Graduate Medical School. These facilities established DES as the first rodent bioimaging center in Singapore.

DES has undergone a difficult and challenging journey to be what it is today, through perseverance and the visionary passion of its Directors. Starting from an empty shell, it has transformed into a full-fledged research institution of international standing. This journey in many ways reflects the development of translational research and the Biomedical initiative the country has embarked on since 2000 — something that the staff can be proud of and should aim to push to further heights. The Department of Experimental Surgery currently offers the requisite equipment support, expert manpower services and laboratories for the conduct of translational research projects that supports 9 research programmes. They are:

- Neurobiology, which assesses cellular behaviour in spinal injuries and develop treatment regiment such as vaccines.
- Myocardial Infarction Studies which use stem cells and engineered skeletal muscle cells to rejuvenate a failing heart.
- Ophthalmology Studies which focus on diseases and treatment in ophthalmology.

- Novel therapies in Diabetes, which evaluates islet cell implants and gene therapy to augment compromised insulin production.
- Trauma Management, which consists of studies on haemorrhagic shock, brain injury and dermal burns.
- Experimental Oncology, which comprises of projects involving therapeutic and diagnostic strategies in cancer using microPET and microCT as analytical tools.
- Transplant Immunology, which focuses on immunologic responses in cellular transplantation.
- Hepatobiliary Disease involving development of biliary stents and models of hepatitis and cirrhosis induced by Hepatitis B infection.
- ADME studies involving pharmacokinetics and toxicology evaluation of drugs.

Core Manpower and Key Institutional Representative

Below (Table 8.1.1) is a list of core manpower:

Table 8.1.1: List of core manpower

Tan Ser Kiat	CEO SGH, appoints SingHealth-IACUC members
Malcolm Paterson	SingHealth Research Secretariat Chairman SGH Division of Research
Pierce Chow	Director DES, First IACUC Chairman Chair IACUC Subcommittee of NACLAR
Bryan Ogden	Institutional Veterinarian, SGH Chair, Emergency Crisis Management Committee
Robert Ng Teck Hin	Senior Manager, DES Member of Training Subcommittee of NACLAR
Fong Yoke Tien	Occupational Health Physician Chairperson Animal Facilities Biosafety Committee
Asliyah Bte Amat	DES Secretary
Inria Kurniawan	DES Research Executive and Training Coordinator
Technical Team (8 personnel)	Headed by Song In Chin and Irene Kee
Scientific Team (4 personnel)	Headed by Senior Scientist, Zhang Kai
Veterinary Support (19 personnel)	Headed by Institutional Veterinarian, Bryan Ogden

Facilities (Major Sites)

DES is divided into five major sites, three of which are located in Block 9, SGH; the fourth is in the Pathology Building, SGH; and the fifth one is off-campus in Sembawang. The area of these sites and their purposes are listed in Table 8.1.2.

Table 8.1.2: Major Sites/Facilities

Description	Area (m^2)	Use
Block 9, Level 2, SGH	554.4	Big animal research and training
Block 9, Level 3, SGH	523.0	Rodent housing and procedures
AH & H, Sembawang	1576.0	Animal breeding and post operative convalescence
Clinical Skills Laboratory	180.0	Surgical skills training
Cadaver Repository	170.0	Anatomical studies

Within these major sites, there are functional rooms equipped with the required materials and equipment. These functional rooms are divided into seven categories, namely:

- Surgery - Operating Theatre Suite, microvascular laboratory, nonhuman primate operating theatre, animal preparation room and necropsy rooms.
- Bioimaging - Micro PET, Micro CT, Bio-luminance, fluoroscopy, gamma camera and X-ray.
- Procedure rooms - Mice procedure, rat procedure, necropsy room, and acclimatization room.
- Support Laboratories - Histology, organic chemistry, molecular biology, biochemistry, microscopy, laser dissection, tissue culture, and virology (BioSafety Level 2, BSL2).
- Skills Training - Clinical skills laboratory, operating theatre suite, and mannequin practice room.
- Animal Holding Unit - Housing for mice, rats, nonhuman primate, rabbits, pigs and sheep, quarantine, and animal BSL2 facility.
- Cadaver repository - Cadaver store (16 cadavers), cadaver dissection and plastination facility and biomechanic laboratory.

Collaborations

Local

1. National Cancer Centre (NCC), SingHealth
2. Singapore Eye Research Institute (SERI), SingHealth
3. National Heart Centre (NHC), SingHealth
4. National Neuroscience Institute (NNI), SingHealth
5. National Dental Centre(NDC), SingHealth
6. National University Hospital

7. National University of Singapore
8. Nanyang Technological University
9. Temasek Polytechnic
10. Defense Science Organisation
11. Institute of Molecular and Cell Biology, A*Star
12. Institute of Bioengineering and Nanotechnology, A*Star
13. Duke-NUS Graduate Medical School

Overseas

1. Universiti Malaysia Sarawak, Sarawak, Malaysia
2. Guangxi Medical University, Guangxi, China
3. Matsumoto Dental University, Japan
4. University of New South Wales, New South Wales, Australia
5. University of Melbourne, Melbourne, Australia
6. Princess Alexandra Hospital, Brisbane, Australia
7. University of Queensland, Brisbane, Australia
8. Wallenburg Laboratory, University Hospital, Malmo, Sweden
9. Tampere University Hospital, Tampere, Finland
10. Institut National de la Santé et de la Recherche Medicalé, France
11. University of Washington, Washington, United States of America
12. University of California, Davis, United States of America
13. Harvard Medical School, Boston, United States of America
14. University of California, Los Angeles, United States of America

Industry

1. pSiOncology Pte Ltd
2. NewBiomed PIKA Pte Ltd
3. Maccine Pte Ltd
4. Vanda Pharmaceutical Inc.
5. Herbal Science Pte Ltd
6. J Morita Manufacturing Co.
7. Dynamed Biotech Pte Ltd
8. Merlin MD Pte Ltd
9. Biosensors
10. Zimmer Pte Ltd
11. Johnson & Johnson Singapore
12. Tyco Healthcare Pte Ltd
13. Karl Storz Endoscopy (S) Pte Ltd
14. Carl Zeiss Pte Ltd
15. Leica Microsystems (SEA) Pte Ltd
16. Mathys Singapore
17. Humedical (S) Pte Ltd

APPENDIX 1

ACCREDITED LABORATORIES FOR IMPORT OF EXPERIMENTAL ANIMALS

Country	Name	Address	Website
Australia	Animal Resource Centre	Murdoch Drive, Murdoch, Murdoch, Western Australia, Australia 6150 (PO Box 1180, Canning Vale, WA 6155)	www.arcwa.wa.gov.au
	Garvan Institute of Medical Research	384, Victoria Street, Darlinghurst, NSW 2010	www.garvan.org.au
	Institute of Medical and Veterinary Science	Veterinary Services Division, 101, Blacks Road, Giles Plain, SA 5086, Australia (PO Box 14, Rundle Mall Post Office, Adelaide, Australia 5000)	www.imvs.sa.gov.au
	The Walter and Eliza Hall Institute of Medical Research	Post Office, Royal Melbourne Hospital, Victoria 3050	www.wehi.edu.au
Austria	Institute of Molecular Pathology	Dr Bohgasse 7, A1030 Vienna	www.imp.univie.ac.at
Canada	Montreal Neurological Institute and Hospital	McGill University. Montreal, Ontorio, Canada, H3A 2B4	www.mni.mcgill.ca
	Ontario Cancer Institute, Princess Margaret Hospital	500, Shebourne Street, Toronto, Ontario, Canada, M4X	
France	IFFA CREDO	69210 St Germaine Sur L'aarbresle, BP 0109-69592, l"Arbresle Cedex	www.criver.com
Germany	Barrier Unit (Barrier 104), Laboratory Animal Facility	University Hospital Eppenorf, Hamburg	www.uke.uni-hamburg.de
	The Institute for Genetics-IMCB (GM Animals)	University of Cologne, Germany, Weyertal 121 D-50931	www.genetik.uni.koeln.de
Greece	Biomedical Sciences Research Centre "Alexander Fleming"	Institute of Immunology, Animal Facility, 14-16 Al, Fleming Street, 16672, Vari-Athens	www.fleming.gr
Hong Kong	Laboratory Animal Services Centre	C-/Shanghai Fraternity Association Research Services Centre, Chinese University of Hong Kong, Shatin, New Territories	www.hku.hk

Japan	Seac Yoshitomi Ltd	955 Koiwai Yoshitomi-cho, chikujo-gun, Fukuoka 971-8550	
	The Institute for Animal Experimentation	Tohoku University School of Medicine, 2-1 Sieryo-cho Aoba-ku, Sendai 980-8575	www.tohoku.ac.jp
	Animal Laboratory for Biomedical Research, Graduated School of Medicine Sciences, Kyushu University	53 Shogoin Kawahara-cho, Sakyo-kyu, Kyoto 606-8597, Japan	www.anim.med.kyoto-u.ac.jp
	Genome Information Research Centre	Osaka University, Yamadaoka 3-1, Suita Osaka, 565-0871	www.gen-info.osaka-u/ac/jp
	CLEA Japan Inc	20-14. Aobadai-2, Meguro-ku, Tokyo, Japan	www.clea-japan.com
	Central Institute for Experimental Animals	1430 Nogawa, Miyamae-shi, Kanagawa	www.rash2.com
	Animal Centre for Biomedical Research, Facility of Medicine, University of Tokyo	Animal Centre for Biomedical Research, Facility of Medicine, University of Tokyo, 7-3-1 Hongo, Bunkyo-ku, Tokyo 113-0033, Japan	www.u-tokyo.ac.jp
	Jichi Medical School	3311-1 Yakushiji, Minanikawachi, Kawachi, Tochigi 32-0498, Japan	www.jichi.ac.jp
	Animal Facility of Centre of Biomedical Research, Graduated School of Medicine Sciences, Kyushu University	3-1-1 Maidashi, Higashi-ku, Fukuoka, Fukuoka 812-8582, Japan	www.med.kyushu-u.ac.jp
	Laboratory Animal Resource Centre, University of Tsukuba	1-1-1 Tennodai, Tsukuba, Ibaraki 305-8575, Japan	www.tsukuba.ac.jp
Indonesia	Primate Research Centre	Bogor Agricultural University, Jalan Lodaya II No. 3, Bogor 16151	www.ipb.ac.id
The Netherlands	Broekman Institute B.V.	Schoolstraat 21, 5711 CP Someren	
	Harlan Netherlands BV	Kreuzelweq 53, PO Box 6174, Netherlands	www.harlan.com
United Kingdom	B & K Universal Ltd	Grimston, Aldbrough, Hull, North Humberside, Hull 4QE	
	Department of Medicine	University of Bristol, Bristol University Medical School, University Walk, Bristol BS8 ITD	www.bris.ac.uk
	Harlan Interfauna Limited	Abbots Ripton Road, Wyton, Huntingdon, Cambs, England, PE 17 2DT	www.harlan.com
	Biomedical Services Department	Royal Liverpool University Hospital, Duncan Building Daulby Street, Liverpool, L7 8XW	www.rlbuht.nhs.uk
	Harlan UK Ltd	Shaw's Farm Blackthorn, Bicester, Oxon, OX6 OTP	www.harlan.com
Canada	John P. Robarts Research Institute	John P. Robarts Research Institute Box 5015, 100 Perth Drive London, Ontario N6A 5K8	www.robarts.ca

USA	Animal Resource Facilities	University of South Carolina, School of Medicine, Graduated Science Research Centre, Suite 102, Columbia, South Carolina 29208	www.med.sc.edu
	Charles River Laboratories	251 Ballardvale Street, Wilmington, MA 01887	www.criver.com
	Charles River Laboratories/ Animal Production Area	National Cancer Institute – Frederick Cancer Research and Development Centre, Building 1021, Frederick, MD 21702	www.criver.com
	Charles River Laboratories (Provisional)	Transgenic Services San Diego, 10792 Roselle St., CA 92121	www.criver.com
	College of Physicians and Surgeons	Columbia University, 701 West 168th Street, NY 10032	www.criver.com
	Taconic	273 Hover Avenue Germanton, NY	www.taconic.com
	Transgenic Animal Barrier Facility (TBAF) – IMCB (GM Mice)	University of Cinninnati, 231, Bethesda Avenue, Cinninati, OH 45267-0571	www.uc.edu
	Department of Animal Care	Oregon Health Sciences University, 3181 SW Sam Jackson Park Road, Portland, Oregon 97201	www.ohsu.edu
	The Jackson Laboratory	600 Main Street, Bar Harbour, Maine, ME 04609	www.jax.org
	Lexicon Genetics Incorporate	4000 Research Forest Drive, The Woodlands, TX 77381	www.laxgen.com
	Roswell Park Cancer Institute	Department of Lab Animal Resources, Elm & Carlton St Buffalo, NY 14263	www.roswellpark.org
	Baylor College of Medicine	Baylor College of Medicine, Houston, Texas	www.bcm.tmc.edu
	Massachusetts General Hospital	Centre for Comparative Resources, Research Affairs Lawrence E., Martin Laboratories, 149 Thirteenth St 149, 08, 424 Charlestown, Massachusetts 02129-2000	www.massgeneral.org
Sweden	Facility of Gnotobiology	Karolinska Institute, von Eulers vag 5, 171 88n Stockholm, Sweden	www.info.ki.se
Switzerland	Swiss Federal Institute of Technology	Zurich	www.epfl.ch

APPENDIX 2

RADIATION SAFETY DATA

Radiation Safety Data

- The Department of Experimental Surgery uses the electronic pocket dosimeter that is suited for measuring the accumulated dose equivalent to X rays of 20 KeV.

- All personnel conducting radioactive procedure or handling animals injected with radionucleotide have to clock and record the radiation exposure dose on a radiation exposure log book.

- Radiation outside the animal cage should not exceed 2 millrem/hr or 20 mSv/hr.

- Used cages with soiled and contaminated bedding should be kept in designated refrigerated containment with 1" thick Perspex container. Length of time to be kept will be equivalent to twice the half life of the radionucleotide use before they can be dispose safely as biohazard waste.

- Radiation Protection (Ionising Radiation) Regulation 2000 stipulates the dose limit for a radiation worker is 20 mSv per year which works out to 35 mSv per week.

- Radiation dose can be averaged out such that the total dose does not exceed 20 mSv/year. For example, a radiation worker may be subjected to say 500 mSv/week for a couple of months (which is over the dose limit of 385 mSv per week). What he can do is to reduce the number of hours of exposure in the next few months such that after averaging out he can keep within the 20 mSv dose limit at the end of the year.

- Acrylic beta syringe shield or a leaded gamma syringe shield should be used for injection of radionucleotide that emits beta or gamma rays respectively.

- Conversion rate: 1 Sv (Sievert) = 100 rem
  Sv = Gy × QF (Quantity Factor is 1 for Beta and Gamma radiation).

APPENDIX 3

ANAESTHESIA FOR LABORATORY ANIMALS

Specific Notes:

1. Ten per cent Lignocaine spray is used as an antivaspasmotic agent to relief laryngeal spasm.
2. Pancuronium (0.02–0.15 mg/kg) is given IV to induce paralysis of the diaphragm during cardiac surgery. It should not be administered until surgical anaesthesia is established.
3. For liver and heart surgeries, Isoflurane is substituted for Halothane, which has depressive effects on myocardium and is hepatotoxic.

Table 1: Anaesthesia for laboratory animals

Animal	Premedication	Induction	Maintenance	Tidal vol	Breath/ min
Mouse (14–20 g)	Atrophine 0.05 mg/kg (SC) 30 mins before induction	Ketamine 50 mg/kg + Diazepam 5 mg/kg (IP)	1.5% Halothane	0.15 ml	180
Rat (250 g)	Atrophine 0.05 mg/kg (SC) 30 mins before induction	Ketamine 50 mg/kg + Diazepam 5 mg/kg (IP)	1.5% Halothane	1.6 ml	90
Hamster (100 g)	Atrophine 0.05 mg/kg (SC)	Ketamine 50 mg/kg + Diazepam 5 mg/kg (IP)	1.5% Halothane	0.8 ml	80
Tree Shrew (200 g)	Atrophine 0.1 mg/kg (SC)	Ketamine 50 mg/kg + Xylazine 10 mg/kg (IP)	1.5% Halothane	1.5 ml	90

Guinea Pig (500 g)	Atrophine 0.05 mg/kg (SC)	Ketamine 50 mg/kg + Xylazine 10 mg/kg (IP)	2% Halothane	2.5 ml	120
Rabbit (2.5–4.0 kg)	-	Ketamine 50 mg/kg + Xylazine 10 mg/kg (IP)	2% Halothane	20 ml	50
Monkey (2–4 kg)	Atrophine 0.05 mg/kg (SC)	Ketamine 25 mg/kg (IM)	2% Halothane	39 ml	38
Pig (30–40 kg)	Atrophine 0.05 mg/kg (SC) Ketamine 15 mg/kg (IM)	Halothane 5% or Pentobarbitone 15 mg/kg (IV)	2–3% Halothane	12 ml/kg	12–15
Sheep (40 kg)	Tetracyline 25 mg/kg (IM) the day before	Xylazine 0.1 mg/kg (IM)	2% Halothane	280–330 ml	20
Dog (15 kg)	Ketamine 20 mg/ml (IM)	Pentobarbitone 25 mg/kg (slowly IV)	1.5% Halothane	250 ml	20

APPENDIX 4

ANALGESICS AND THERAPEUTICS FOR LABORATORY ANIMALS

Primates

Table 1: Common analgesic drugs for macaques

Analgesic	Dosage/Route	Duration
Buprenorphine	0.01 mg/kg IN, IV	6 to 8 hours
Morphine	1 to 2 mg/kg SC, IM	4 hours
Oxymorphone	0.15 mg/kg IM, SC, IV	4 to 6 hours
Fentanyl	5 to 10 mcg/kg or 10 to 25 mcg/kg/hr IV infusion	Intraoperatively
Aspirin	125 mg/5 kg rectal suppository	Less 24 hours
Keterolac	15 to 30 mg/kg IM	

Table 2: Emergency drugs

Drug	Indication	Dosage/Route
Atropine	Bradycardia	0.05 mg/kg IV
Dopamine	Bradycardia after arrest 10 mcg/kg/min IV (must be above 3 kg)	10 mcg/kg/min IV (must be above 3 kg)
Doxapram	Respiratory arrest	2 mg/kg IV
Epinephrine	Cardiac arrest	0.2 to 0.4 mg/kg diluted in 5 ml sterile water; give intratracheally >3 kg or 1:10,000 dilution 0.5 to 1.0 ml IV
Ephedrine	Hypotension	1.25 to 2.5 mg/kg IV
Furosemide	Pulmonary edema	1 to 2 mg/kg IV
Lidocaine	Premature ventricular contraction	1 to 2 mg/kg IV
Nalaxone	Opiod reversal	0.1 to 0.2 mcg IV, repeat as needed
Norepinephrine	Hypotension	0.05 to 0.1 mcg/kg/min, IV infusion
Phenylephrine	Hypotension	1 to 2 mcg/kg IV bolus, then 0.5 to 1.0 mcg/kg/min, IV infusion

Swine

Table 3: Anesthetic/Analgesic drugs for swine

Drug	Dose (mg/kg)	Route	Duration	Comments
Acepromazine	0.11 to 2.2	IM	8 to 12 hours	Moderate sedation; no analgesia or anesthesia
Atipamazole	0.24	IM/IV	Alpha-2 antagonist	
Atropine	0.01 to 0.05	SC/IM/IV	Anticholinergic	
Azaperone	1.0 to 8.0 IM	IM	1 to 2 hours	Moderate to deep sedation; no analgesia
Buprenorphine	0.20 to 0.30/0.005 to 0.10/0.1 to 0.12	SC/IM/IV	6 to 12 hours	Good analgesia; some sedation
Butorphanol	0.1 to 0.4	SC/IM/IV	4 to 6 hours	Potent analgesic
Caprofen	0.5 to 4.0	SC/IM/IV	24 hours	Can be administered preoperatively
Diazepam	0.5 to 10/0.44 to 2.0	IM/IV	2 to 4 hours	Sedation; muscle relaxation; no analgesia
Doxapram	0.5 to 1.0/5 to 10 ug/kg/hr	IV/IV/ infusion	15 to 20 minutes	Analeptic
Fentanyl	0.02 to 0.15	IM/IV	2 hours	Potent analgesic; depressed ventilation; not as post-op analgesic
Flumezenil	At a dose of 1 part Flumazenil to 13 parts benzodiazepine			Benzodiazepine antagonist
Flunixin	1.0 to 2.2	SC/IM/IV	24 hours	Post-op analgesic
Glycopyrolate	0.004 to 0.01	IM	30 minutes	Anticholinergic
Ketamine	2 to 33 mg	IM/IV	Sedation; mobilization but poor muscle relaxation, inadequate analgesia; not as sole agent	
Ketoprofen	1.0 to 3.0	SC/IM/IV	24 hours	Potent analgesic; post-op and chronic pain

Mice

Table 4: Analgesics and corresponding doses for mice

Analgesic	Doses
Buprenorphine	1.0 to 2.0 mg/kg SC given 12 hourly
Butorphanol	1 to 5 mg/kg SC for 4 hours of analgesia
Codeine	60 to 90 mg/kg orally or 20 mg/kg SC for 4 hours of analgesia
Morphine	2 to 5 mg/kg SC for 2 to 4 hours of analgesia
Nalbuphine	4 to 8 mg/kg IM for 4 hours of analgesia
Flunixin	2.5 mg/kg SC or IM lasts 12 hours
Ibuprofen	30 mg/kg orally lasts 4 hours
Diclofenac	8 mg/kg orally
Paracetamol	200 mg/kg orally lasts 4 hours
Aspirin	120 mg/kg orally lasts 4 hours
Phenylbutazone	30 mg/kg orally

Rats

Table 5: Analgesics and corresponding doses for rats

Analgesic	Doses
Buprenorphine	0.1 mg/kg SC given every 8 to 12 hours
Butorphanol	0.5 to 2.0 mg/kg SC for 4 hours analgesia
Codeine	60 mg/kg SC for 4 hours analgesia
Morphine	2 to 5 mg/kg SC for 2 to 4 hours analgesia
Nalbuphine	1 to 2 mg/kg IM for 3 hours analgesia
Pentazocine	10 mg/kg SC for 3 to 4 hours analgesia
Pethidine	10 to 20 mg/kg SC, IM for 2 to 3 hours analgesia
Carprofen	10 mg/kg oral or 5 mg/kg SC twice a day
Flunixin	2.5 mg/kg once daily
Phenylbutazone	20 mg/kg orally
Diclofenac	10 mg/kg oral
Aspirin	100 mg/kg orally lasts for 4 hours
Ibuprofen	15 mg/kg orally lasts 4 hours
Paracetamol	100 to 300 mg/kg orally lasts 4 hours

* If the rat is eating, then continued analgesia maybe provided by mixing the required amount of the drug in jelly so it is taken orally.

Rabbits

Table 6: Analgesics and dosage/route for rabbits

Analgesic	Dosage/route	Degree of pain, duration
Aspirin	100 mg/kg oral in solution	Mild to moderate, 4 hours
Butorphanol tartarate	0.1 to 1.5 mg/kg, IV or 1.0 to 7.5 mg/kg SC or IM	Mild to moderate, 4 hours
Buprenorphine	0.01 to 0.05 mg/kg, SC or IV	Severe 6 to 12 hours
Morphine	2.5 mg/kg, SC	Severe 2 to 4 hours

APPENDIX **5**

PASSIVELY TRANSMITTED ZOONOTIC ORGANISMS

Common Name	**Organism**	**Animals of Concern**	**Risk/Concern**	**Symptoms of Infection in Humans**
Brucellosis	*Brucella* sp.	Dogs, cattle, sheep, swine and goats	Low/ moderate-high	Gradual onset, undulating fever, chills, sweats, headache, myalgia, fatigue, backache, weakness, malaise, weight loss, extended convalescence; may be chronic, with stress related relapses, complications may include emotional and cardiac symptoms, arthritis.
Colibacillosis	*Escherichia coli*	Vertebrates	Low/moderate	Pneumonia, diarrhoea, urinary tract disease.
Hantaan virus (Korean haemorrhagic fever)	*Hantaan* virus	Wild or unscreened rodents	Low/high	Subtle onset; malaise, fever with neurologic disturbances, common renal shutdown, headache, tremors of tongue and extremities, shock. **30–40 % fatality rate!**
Lymphocytic choriomeningitis	*LCM* virus (*arenavirus*)	Rodents	Low/high	Fever, myalgia, malaise, occasional stiff neck, headache, sleepiness, unusual skin sensations (paresthesia), paralysis, usually self-limiting. **Some fatalities!**
Plague (bubonic and pneumonic)	*Yersinia pestis*	Ground squirrels, wild-caught rodents	Low/high	*Bubonic* — fever, chills, nausea, diarrhoea or constipation, headache, meningitis, tachycardia, coma, regional lymphadenopathy. **60 % fatality rate if untreated.** *Penumonic* — cough and dyspnoea with mucoid to bring red sputum; may progress to septicaemic form, with vascular collapse, haemorrhagic rash. **95 % fatality rate in these two forms if untreated.**

Pneumocystis pneumonia	*Pneumocystis carinii*	Rodents, guinea pigs, rabbits, dogs, cats, cattle, sheep, swine, monkeys	High in immuno-compromised patients	Generally seen only in those with serious underlying disease or suppressed immune system; pneumonia, dyspnoea, nonproductive cough, moderate fever, tachypnea, cyanosis.
Q-fever	*Coxiella burnetti*	Sheep, cattle, goats	Moderate/ moderate	Sudden onset of fever, retrobulbar of frontal headache, chills, myalgia, sweating, weakness, malaise, pneumonitis, endocarditis, hepatitis.
Ringworm	*Microsporum* and *Trichophyton* spp.	Cats, rabbits, dogs	High/low	Generally, scaling, hair loss or breakage; occasional itching; less frequently, erythema, induration, crusting, suppuration.
Salmonellosis	*Salmonella* spp.	All	Low/moderate	Diarrhoea, vomiting, low-grade fever, may progress to dehydration, prostration, death; septic syndrome has high-spiking fever, septicaemia, splenomegaly, headache.
Simian haemorrhagic fevers (Ebola, Marburg)	*Rhabdovirus*	Nonhuman primates	Low/very high	Fever, malaise, headache,myalgia, vomiting, conjunctival infection, diarrhoea, sore throat, haemorrhages. **High fatality rate!**
Toxoplasmosis	*Toxoplasma gondii*	Cats	Moderate/ moderate	Usually, lymphadenopathy, fever, headache, myalgia, stiff neck, anorexia; occasionally, arthalgia, maculapapular rash, confusion.
Tuberculosis	*Mycobacterium* spp.	Cattle, birds, nonhuman primates, humans	Moderate to high/moderate to high	Pulmonary — productive cough, fever, weight loss, fatigue, night sweats, chest pain, haemoptysis.
Yaba virus	*Pox* virus	Nonhuman primates	Moderate/ moderate	Papulae develop to subcutaneous tumours on limbs, hands, feet, face, ears; regional lymphadenopathy.

REFERENCES

Chapter 1.1: The Rationale for the Use of Animal Models in Biomedical Research

Mack P. (1994) *Clinician's Guide to Experimental Surgery*. Image Medicus, Singapore.

Lumley J. S. P., Green C. J. *et al.* (1990) *Essentials of Experimental Surgery*. Butterworths, London.

Calman K. C. (1989) The choice of an experimental model. In *Principles of Surgical Research*. Eds. Mathie R. T., Taylor K. M. and Calnan J. S. Butterworths, London.

Heitman E., Reiser S. J. and Bulger R. E. (1996) Legal, ethical and educational problems in experimental surgery. In *Essentials of Experimental Surgery: Gastroenterology*. Eds. Jensen S. L., Gregersen H. *et al.* Harwoods Academic, Amsterdam.

Isselhard W. H. and Kushe J. (1986) Animal experimentation. In *Principles and Practice of Research: Strategies for Surgical Investigators*. Eds. Troidl H., Spitzer W. O. *et al.* Springer-Verlag, Berlin.

Nuffield Council on Bioethics (2005) *The Ethics of Research Using Animals*. http://www.nuffieldbioethics.org

Chapter 1.3: Nonhuman Primates as Models in Biomedical Research

Rhesus Monkey Demands in Biomedical Research: A Workshop Report. Based on a workshop held April 19–20, 2002 at the National Academy of Sciences, Washington DC.

Wagner J. D., Cline M., Shadoan M. K. *et al.* (2001) Naturally occurring and experimental diabetes in Cynomolgous monkeys: a comparison of carbohydrate and lipid metabolism and islet pathology. *Toxicol. Pathol.* **29**(1): 142–148.

Walsh G. P., Esterlina V. T., Cruz E. C. D. *et al.* (1996) The Philippine cynomolgous monkey (*Macaca fasicularis*) provides a new nonhuman primate model of tuberculosis that resembles human disease. *Nat. Med.* **2**(4): 430–436.

Chapter 2.5: Use of Statistics as Determinant for Number of Animals Used

Chow S. C. (1998) Sample size calculation. In *Introduction to Clinical Trials*. Eds. Karlberg J. and Tsang K. University of Hong Kong, Hong Kong, pp. 267–286.

Festing M. F. W., Overend P., Das R. G., Borja M. C. and Berdoy M. (2002) *The Design of Animal Experiments: Reducing the Use of Animals in Research Through Better Experimental Design*. Laboratory Animal Handbooks No. 14, Royal Society of Medicine Press.

Machin D., Campbell M. J., Fayers P. M. and Pinol A. P. Y. (1997) *Sample Size Tables for Clinical Studies*, 2nd edn. Blackwell Science Ltd.

Chapter 4.1: Bioimaging in Animals

Carmichael S. T., Tatsukawa K., Katsman D. *et al.* (2004) Evolution of diaschisis in a focal stroke model. *Stroke* **35**: 758–763.

Cherry S. R., Shao Y., Silverman R. W. *et al.* (1997) MicroPET: a high resolution PET scanner for imaging small animals. *IEEE Trans. Nucl. Sci.* **44**: 1161–1166.

Couturier O., Luxen A., Chatal J.-F. *et al.* (2004) Fluorinated tracers for imaging cancer with postitron emission tomography. *Eur. J. Nucl. Med. Mol. Imaging* **31**: 1182–1206.

Elsinga P. H. (2002) Radiopharmaceutical chemistry for positron emission tomography. *Methods* **27**: 208–217.

Fowler J. S. and Ding Y.-S. (2002) Chemistry. In *Principles and Practice of Positron Emission Tomography*. Eds. Wahl R. L. and Buchanan J. W. Lippincott Williams and Wilkins, Philadelphia, pp. 6–47.

Fueger B. J., Czernin J., Hildebrandt I. *et al.* (2006) Impact of animal handling on the results of 18F-FDG PET studies in mice. *J. Nucl. Med.* **47**: 999–1006.

Huber J. S. and Moses W. W. (1999) Conceptual design of a high-sensitivity small animal PET camera with 4P coverage. *IEEE Trans. Nucl. Sci.* **46**: 498–502.

Ido T., Wan C.-N., Casella V. *et al.* (1978) Labeled 2-deoxy-D-glucose analogues. ^{18}F-labeled 2-deoxy-2-fluoro-D-glucose, 2-deoxy-2-fluoro-D-mannose and 14C-2-deoxy-2-fluoro-D-glucose. *J. Labeled Comp. Radiopharm.* **14**: 171–183.

Kornblum H. I., Araujo D. M., Annala A. J. *et al.* (2000) *In vivo* imaging of neuronal activation and plasticity in the rate brain by high resolution positron emission tomorgraphy (MicroPET). *Nat. Biotechnol.* **18**: 655–660.

Lu Y., Dang H., Middleton B. *et al.* (2006) Non-invasive imaging of islet grafts using positron-emission tomography. *Proc. Natl. Acad. Sci. USA* **103**: 11294–11299.

Okuma T., Matsuoka T., Okamura T. *et al.* (2006) 18F-FDG small-animal PET for monitoring the therapeutic effect of CT-guided radiofrequency ablation on implanted VX2 lung tumors in rabbits. *J. Nucl. Med.* **47**: 1351–1358.

Oyama N., Ponde D. E., Dence C. *et al.* (2004) Monitoring of therapy in androgen-dependent prostate tumor model by measuring tumor proliferation. *J. Nucl. Med.* **45**: 519–525.

Phelps M. E. (2000) Positron emission tomography provides molecular imaging of biological processes. *Proc. Natl. Acad. Sci. USA* **97**: 9226–9233.

Tewson T. J. and Krohn K. A. (1998) PET radiopharmaceuticals: state-of-the-art and future prospects. *Semin. Nucl. Med.* **28**: 221–234.

Welch M. J. and Redvanly C. S., Eds. (2003) *Handbook of Radiopharmaceuticals: Radiochemistry and Application*. John Wiley & Sons, West Sussex, UK.

Zhang Y., Saylor M., Wen S. *et al.* (2006) Longitudinally quantitative 2-deoxy-2[(18)F]fluoro-D-glucose micro positron emission tomography imaging for efficacy of new anticancer drugs: a case study with bortezomib in prostate cancer murine model. *Mol. Imaging Biol.* **8**(5): 300–308.

Chapter 4.4: Animal Cell Culture

Ryan J. A. (2003) *Introduction to Animal Cell Culture*. Technical Bulletin.

Chapter 4.5: Application of Microsurgical Techniques in Animal Research

Barker J. H. and Ryan T. J. (1998) Skin microcirculation. Clinically applied microcirculation research: 317.

Chang S. Y., Chen H. C. and Tang Y. (1999) Prefabrication of jejunum for challenging reconstruction of cervical esophagus. *Plast. Reconstr. Surg.* **104**: 2112–2115.

Chen H. C., Tan B. K., Cheng M. H. *et al.* (2002) Behavior of free jejunal flaps after early disruption of blood supply. *Ann. Thorac. Surg.* **73**: 987–989.

Dolderer J. H., Findlay M. and Cooper-White J. (2003) *In-vivo* generation of 3-dimensional vascularized tissue constructs. *ANZ J. Surg.* **73**: 250.

Erol O. O. and Spira M. (1980) Secondary musculocutaneous flap: an experimental study. *Plast. Reconstr. Surg.* **65**: 277–282.

Erol O. O. (1976) The transformation of a free skin graft into a vascularized pedicled flap. *Plast. Reconstr. Surg.* **58**: 470–477.

Erol O. O. and Spira M. (1980) New capillary bed formation with a surgically constructed arteriovenous fistula. *Plast. Reconstr. Surg.* **66**: 109–115.

Findlay M., Dolderer J. and Cooper-White J. (2003) Creating large amounts of tissue for reconstructive surgery — a porcine model. *ANZ J. Surg.* **73**: 240.

Fisher J. and Wood M. (1987) Experimental comparison of bone revascularisation by musculocutaneous and cutaneous flaps. *Plast. Reconstr. Surg.* **79**: 81–90.

Frank J. M. (1995) Angiogenesis. In *Clinically Applied Microcirculation Research.* Eds. Barker J. H., Anderson G. L. and Menger M. D. CRC Press, Boca Raton, FL, USA, pp. 375–389.

Hill M. A. and Meininger G. A. (1995) Skeletal muscle microcirculation. In *Clinically Applied Microcirculation Research.* Eds. Barker J. H., Anderson G. L. and Menger M. D. CRC Press, Boca Raton, FL, USA, pp. 297–314.

Khouri R. K., Upton J. and Shaw W. W. (1991) Prefabrication of composite free flaps through staged microvascular transfer: an experimental and clinical study. *Plast. Reconstr. Surg.* **87**: 108–115.

Morrison W. A., Dvir E., Doi K. *et al.* (1990) Prefabrication of thin transferable axial-pattern skin flaps: an experimental study in rabbits. *Br. J. Plast. Surg.* **43**: 645–654.

Shintomi Y. and Ohura T. (1982) The use of muscle vascularized pedicle flaps. *Plast. Reconstr. Surg.* **70**: 725–735.

Tan B. K., Chen H. C., Wei F. C. *et al.* (2002) Formation of independently revascularized bowel segments using the rectus abdominis muscle flap: a rat model for jejunal prefabrication. *Plast. Reconstr. Surg.* **109**(2): 655–663.

Tham C., Song C. and Ng R. (2007) Free flap prefabrication with a bioengineered acellular dermal replacement (Integra®) in a rat model. *Plast. Reconstr. Surg.* (submitted).

Chapter 5.1: Species Specific Caging Configuration and Design

Guidelines on the Care and Use of Animals for Scientific Purposes — National Advisory Committee for Laboratory Animal Research (2004).

Laboratory Animal Technician Training Manual. American Association for Laboratory Animal Science (2005).

Chapter 6.3: Radiation Safety Awareness in Animal Research

SGH Radioactive Waste Manual, document number: 98000-E2P-010.

SGH Radiation Safety Manual, document number: 98000-E2P-106.

The Radiation Protection Regulations, 2000.

The Radiation Protection Act (Chapter 262), 1991.

Applied Radiation Biology and Protection, Robert Granier, Ellis Horwood (1990), ISBN 0-13-039991-4.

Introductory Physics of Nuclear Medicine, Ramesh Chandra, Lea & Febiger (1992), ISBN 0-8121-1442-6.

Radiation Biology, Alison P. Casarett, Prentice-Hall, Inc. (1968), Library of Congress catalog card number 68-22702.

INDEX AND KEYWORDS